FLORA OF TROPICAL EAST AFRICA

CAMPANULACEAE

MATS THULIN

(University of Uppsala)

Annual or perennial herbs, subshrubs, or rarely small shrubs, laticiferous. Leaves alternate, rarely opposite, simple, entire, dentate to incised or rarely variously lobed, exstipulate. Inflorescences generally cymose, panicle-, raceme-, spike- or head-like, or flowers solitary. Flowers bisexual, usually protandrous, regular, (3–)5(–10)-merous, mostly with a bract and 2 bracteoles. Calyx ± adnate to the ovary; lobes usually free, persistent, valvate. Petals connate to various degrees, sometimes almost free, valvate in bud. Stamens alternating with the corolla-lobes, free or rarely adnate to the corolla; anthers very rarely entirely or partly connate, introrse; filaments usually dilated at the base. Ovary ± inferior, rarely superior, 2–10-locular; ovules few–many, anatropous, on axile placentas; style 1, furnished with pollen-collecting hairs on and usually below the style-lobes. Fruit capsular, variously dehiscing by apical or lateral valves or pores, or ± baccate. Seeds 1–many, albuminous; embryo straight, terete.

About 35 genera and some 700 species, especially well represented in the Mediterranean region and South Africa, but relatively sparsely developed in the tropics.

The family is taken here in a restricted sense, excluding the closely related Lobeliaceae which is sometimes regarded as a subfamily of Campanulaceae. Lobeliaceae is a generally more specialized group, differing, for example, by having zygomorphic flowers, usually connate anthers and racemose inflorescences, and pollen grains with differences in shape and exine structures. Sphenocleaceae is sometimes also included in Campanulaceae.

1. Corolla ± 5 cm. long, orange or red;
   flowers 6-merous; fruit baccate . . . 1. **Canarina**
   Corolla less than 3 cm. long, never orange or red;
   flowers (3–)5-merous; fruit capsular 2
2. Capsule dehiscing by apical valves (above the insertion of the calyx-lobes, (fig. 2/I–M, p. 6) . . 2. **Wahlenbergia**
   Capsule dehiscing laterally (fig. 7/E, p. 36) or indehiscent 3
3. Capsule indehiscent, tardily opening by the irregular decomposition of the pericarp between the persistent lateral nerves; seeds sometimes with hair-like projections . . . . . 3. **Gunillaea**
   Capsule dehiscing by lateral pores (basal in the Flora area); seeds without hair-like projections . 4. **Campanula**

1. CANARINA

L., Mant. Pl. Alt.: 148, 588 (1771); Hedb. (et al.) in Sv. Bot. Tidskr. 55: 17–62 (1961), *nom. conserv. prop.*

Mindium Adans., Fam. Pl. 2: 134 (1763), *nom. rejic. prop.*

Pernetya Scop., Introd. Hist. Nat.: 150 (1777), *non Pernettya* Gaudich. (1829)

Glabrous terrestrial or epiphytic perennial herbs containing abundant white latex. Roots thickened, fleshy. Stems herbaceous, terete and hollow, di- or trichotomously branched from the leaf-axils; most leaf-axils also produce small, usually rudimentary, accessory shoots. Leaves opposite or ternate, petiolate. Flowers solitary in dichasial forks or terminal, large, pendent, (5–)6(–7)-merous throughout. Calyx-lobes entire or sometimes dentate, erect, spreading or reflexed. Corolla funnel-shaped or campanulate with short lobes. Filament-bases almost linear to broad, shield-like. Ovary inferior; style shorter than the corolla, markedly thickened towards the apex, with short style-lobes. Fruit baccate, with persistent calyx. Seeds numerous; testa finely pitted or striate.

Three species, two in tropical Africa, one in the Canary Is. The latter, *C. canariensis* (L.) Vatke, is a common greenhouse plant.

Petioles of about the same length as the leaf-blades; pedicels and petioles usually coiled; calyx-lobes entirely united in bud; hypanthium cupular . .   1. *C. abyssinica*

Petioles less than half the length of the leaf-blades; pedicels and petioles not coiled; calyx-lobes free in bud; hypanthium obconical   .   .   .   .   2. *C. eminii*

1. **C. abyssinica** *Engl.* in E.J. 32: 116 (1902) & V.E. 1: 145, fig. 116 (1910); W.F.K.: 85, excl. fig. 63 (1948); Hedb. (et al.) in Sv. Bot. Tidskr. 55: 52, t. 3, fig. 5, 8 (1961); E.P.A.: 1053 (1965); U.K.W.F.: 509, 510 (fig.) (1974). Types: Ethiopia, Galla, *Ellenbeck* 1315 & Shoa, *Ellenbeck* 1568 (B, syn. †); Ethiopia, Shoa, *Ambjörn* 37 (UPS, neo. !, S, iso. !)

Terrestrial herb with tuberous root. Stem much branched at the base, sparsely dichotomously branched in its upper part, up to several m. long, climbing by means of the coiling petioles and pedicels. Petioles of about the same length as the leaf-blades; leaf-blades triangular to pentagonal, 2·5–7·5 cm. long, 1·5–6·5 cm. wide, cordate, acute, obtusely crenate to sharply dentate. Hypanthium broad, cupular, terete. Calyx-lobes 2–3·6 cm. long, 0·5–1 cm. wide, acute, entire, erect, entirely fused in bud forming a cup-like cover, in open flowers usually fused for a few mm. from the base. Corolla 4–7·5 cm. long, funnel-shaped to almost cylindrical, orange-red with darker venation, often lighter inside. Stamens with broad shield-like filament-bases; anthers 9–15 mm. long. Fruit almost spherical, yellow or reddish. Seeds ovate to elliptic, 1·5–2·1 mm. long, 1·0–1·3 mm. wide, dark brown with pitted surface.

UGANDA. Mbale District: Sipi, 31 Aug. 1932, *A. S. Thomas* 426! & Bulago, 29 July 1917, *Snowden* 516! & Elgon, Kapchorwa, 7 Sept. 1954, *Lind* 262!

KENYA. Northern Frontier Province: Lorogi Plateau, 15 June 1959, *Kerfoot* 1051!; Ravine District: W. Lembus Forest, *R. M. Graham* 1003 in *F.D.* 3509!; Nakuru District: Elburgon, 18 Aug. 1951, *Cooper* in *Bally* 8035!

TANZANIA. Arusha District: Ngongongare, 27 June 1955, *Willan* 247!; Buha District: Gwanumpu, 31 Jan. 1974, *Ebbels* 8878!

DISTR. U1, 3; K1–5; T2, 4; Ethiopia and S. Sudan

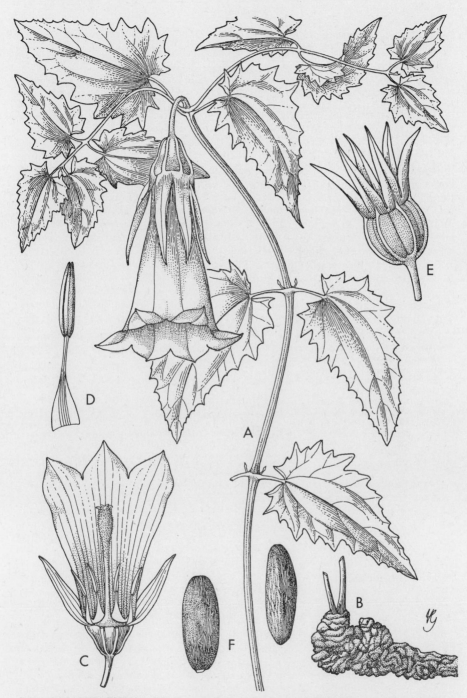

FIG. 1. *CANARINA EMINII*—**A,** habit, × ⅔; **B,** part of root, × ⅔; **C,** flower with two stamens and calyx-lobes and three petals removed, × ⅔; **D,** stamen, × 2; **E,** fruit, × ⅔; **F,** seeds, × 12. A, from *Chojnacki* 8893 & *Norman* 223; B, from *Mooney* 7109; C, D, from *Hedberg* 158; E, F, from *Procter* 2298 & *Norman* 223.

HAB. Shady places in upland wooded grassland, forest clearings, along streams, sometimes on old termite hills; 1350–2300 m.

SYN. *C. abyssinica* Engl. var. *umbrosa* Engl. in E.J. 32: 116 (1902). Type: Ethiopia,
Sidamo, *Neumann* 97 (B, holo. †)

2. C. eminii *Schweinf.* in Sitz. Ges. Naturf. Freunde Berlin 1892: 173

(1892); P.O.A. C: 400, t. 36 (1895); B. L. Burtt in Bot. Mag., t. 9531, fig.
A–D (1938); F.P.N.A. 2: 402 (1947); W.F.K.: 85, fig. 63 (erroneously named
C. abyssinica) (1948); F.P.S. 3: 70 (1956); Hedb. (et al.) in Sv. Bot. Tidskr.
55: 54, t. 4, fig. 6, 9, 13 (1961); E.P.A.: 1053 (1965); U.K.W.F.: 509, 510
(fig.) (1974). Type: Zaire, Ruwenzori, *Stuhlmann* (B, holo. †); Zaire, Ruwenzori, *Humbert* 8818 (P, neo. !, B, BR, US, iso.)

Epiphytic or terrestrial usually glaucous herb. Root thick, often with a
corky surface layer. Stems erect and scandent or pendent, up to several m.
long, dichotomously branched, usually with a fine purplish mottling. Petioles
less than half the length of the leaf-blades, not coiled; leaf-blades ± triangular to ovate, 2·5–10 cm. long, 1·5–9 cm. wide, acute, with cordate to
truncate base, obtusely to sharply dentate to doubly dentate or doubly
serrate. Hypanthium obconical, distinctly 6-ribbed, with the ribs projecting
into the calyx-lobes. Calyx-lobes 1·9–3·8 cm. long, 0·5–1 cm. wide, acute to
acuminate, free, entire, erect or spreading. Corolla 4·3–7·6 cm. long, funnel-
shaped, orange to orange-red with darker venation. Stamens with broad
shield-like filament-bases; anthers 5·5–10 mm. long. Seeds elliptic-oblong,
2·0–2·6 mm long, 0·6–0·8 mm. wide, dark brown, finely striate. Fig. 1.

UGANDA. Acholi District: Imatong Mts., Lomwaga Mt., 5 Apr. 1945, *Greenway &
Hummel* 7284! & Imatong Mts., Langia Mts., Apr. 1943, *Purseglove* 1433!; Mbale
District: Bulago, 28 Aug. 1932, *A. S. Thomas* 340!
KENYA. Elgon, E. slope above Tweedie's saw-mill, 25 Feb. 1948, *Hedberg* 158!; Elgeyo
District: Cherangani Hills, Kaibwibich, Aug. 1968, *Thulin & Tidigs* 59!; Kericho
District: SW. Mau Forest, Sambret to Timbilil, Sept. 1961, *Kerfoot* 2818!
TANZANIA. Rungwe District: Undali [Bundali], 26 Feb. 1933, *R. M. Davies* 902! &
N. slope of Rungwe Mt., 7 Feb. 1961, *Richards* 14274! & near Kiwira Forest Station,
Jan. 1963, *Procter* 2298!
DISTR. U1–3; K2–5; T7; Ethiopia, S. Sudan, E. Zaire, Rwanda, Burundi and Malawi
HAB. Upland or riverine forest, epiphytic or among rocks; 1600–3200 m.

SYN. *C. elegantissima* T. C. E. Fries in N.B.G.B. 8: 392, fig. 1 (1923). Type: Kenya,
Mt. Aberdare, *R. E. & T. C. E. Fries* 2567 (UPS, holo. !, K, S, iso. !)
C. eminii Schweinf. var. *elgonensis* T. C. E. Fries in N.B.G.B. 8: 395, fig. 2 (1923).
Type: Kenya, Elgon, *Lindblom* (S, holo. !)

2. WAHLENBERGIA

Roth, Nov. Pl. Sp.: 399 (1821); v. Brehm. in E.J. 53: 9–143 (1915); Thulin
in Symb. Bot. Upsal. 21(1) (1975), *nom. conserv.*
Lightfootia L'Hérit., Sert. Angl.: 4, t. 4 (1789), *non* Swartz (1788), *nec*
Schreb. (1789), *nom. illegit.*
Cervicina Delile, Fl. d'Egypte: 7, Atlas t. 5/2 (1813), *nom. rejic.*
Cephalostigma A. DC., Monogr. Camp.: 117 (1830)

Annual or perennial herbs, subshrubs or small shrubs. Leaves alternate
or rarely opposite, mostly sessile, simple, entire, dentate or rarely lobed or
incised (but not in East Africa). Inflorescences panicle-, raceme-, spike- or
head-like, or flowers solitary. Calyx-lobes (3–)5. Corolla ± deeply (3–)5-
lobed, or split almost to the base, ± pubescent inside near the base, rarely
with long slender hairs in the corolla-tube (but not in East Africa). Stamens
(3–)5, free; filament-bases variously dilated or linear, usually ciliate. Ovary
subinferior to rarely subsuperior, 2–5-locular; ovules many; style shorter or
longer than the corolla, the upper part with pollen-collecting hairs, glabrous

or hairy below, eglandular or with small glands present at or near the base of the 2–5 style-lobes. Capsule loculicidal, dehiscing by as many apical valves as there are loculi in the ovary. Seeds numerous, ± elliptic in outline; testa smooth or variously reticulate.

Some 200 species, mainly distributed in the southern hemisphere, especially abundant in South Africa.

In keys and descriptions the style is often said to be hairy or hairy below. This does not include the pollen-collecting hairs that are always present on and usually somewhat below the style-lobes (however, they disappear by invagination after anthesis), but concerns the normal hairs often present further down on the style.

When the testa is said to be smooth or almost smooth there are no distinct surface structures to be seen under an ordinary stereo microscope.

1. Corolla-tube $\frac{1}{5}$–$\frac{3}{4}$ the length of the corolla, at least
    1 mm. long 2
   Corolla split almost to the base, the tube usually
    considerably less than 1 mm. long 10
2. Style with tiny glands at the base of the lobes
    (fig. 3/B, p. 10) 3
   Style eglandular 6
3. Leaves few and inconspicuous, usually less than
    7 mm. long and 2 mm. wide; gynoecium 2-
    merous . . . . 2. *W. virgata*
   Leaves ± numerous and larger; gynoecium 3-
    merous 4
4. Testa reticulate (fig. 2/N, U); corolla (5–)8–16 mm.
    long; filament-bases broadly dilated (fig. 2/A);
    inflorescence lax, with pedicels up to 10–25 mm.
    long . . . . 1. *W. undulata*
   Testa smooth; corolla up to 6·5 mm. long; fila-
    ment-bases linear or narrowly dilated; in-
    florescences various 5
5. Inflorescence contracted, with subsessile flowers . 3. *W. songeana*
   Inflorescence lax, with distinctly pedicelled flowers 4. *W. lobelioides*
6. Gynoecium 2-merous; capsule (fig. 2/I) narrowly
    obconical or cylindrical, up to 17 mm. long . 5. *W. silenoides*
   Gynoecium 3-merous; capsule ± obconical or
    hemispherical, shorter 7
7. Calyx-lobes and leaves denticulate; corolla-tube
    $\frac{1}{5}$–$\frac{1}{4}$ the length of the corolla, 1–2(–3) mm. long;
    testa almost smooth . . . . . 9. *W. denticulata*
   Calyx-lobes entire or almost so; leaves various;
    corolla-tube $\frac{1}{2}$–$\frac{3}{4}$ the length of the corolla, more
    than 2·5 mm. long; testa ± reticulate 8
8. Leaves linear; margin not cartilaginous, entire or
    sparsely denticulate; corolla blue; testa faintly
    reticulate . . . . . . 8. *W. capillacea*
   Leaves (linear–) oblanceolate to obovate; margin
    cartilaginous, ± undulate-dentate; corolla ±
    pale with dark venation; testa regularly reticu-
    late (fig. 2/L, M) 9
9. Plants 10–50 cm. tall, with leaves scattered along
    the stem or ± crowded towards the base; flowers
    in a lax inflorescence; pedicels 2–25 mm. long . 6. *W. krebsii*
   Plants dwarf, with leaves ± rosulate; flowers

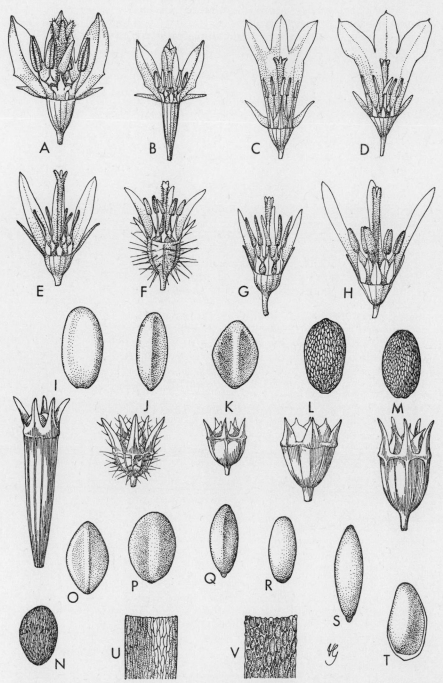

FIG. 2. Flowers, capsules and seeds of *Wahlenbergia* species. **A–H**, flowers with two petals and calyx-lobes removed—**A**, *W. undulata*, × 3; **B**, *W. silenoides*, ×3; **C**, *W. krebsii* subsp. *arguta*, ×3; **D**, *W. capillacea* subsp. *tenuior*, ×3; **E**, *W. subaphylla* subsp. *thesioides*, × 4; **F**, *W. perrottetii*, × 8; **G**, *W. abyssinica* subsp. *abyssinica*, × 4; **H**, *W. denticulata*, × 4. **I–M**, capsules and seeds—**I**, *W. silenoides*, ×3 and × 30; **J**, *W. pulchella* subsp. *pedicellata*, × 4 and × 30; **K**, *W. hookeri*, ×4 and × 30; **L**, *W. pusilla*, × 4 and × 30; **M**, *W. krebsii* subsp. *arguta*, × 3 and × 30. **N–T**, seeds, ×30—**N**, *W. undulata*; **O**, *W. perrottetii*; **P, Q**, *W. abyssinica* subsp. *abyssinica*, two forms; **R**, *W. napiformis*; **S**, *W. capillacea* subsp. *tenuior*; **T**, *W. denticulata*. **U, V**, details of seed surface, × 90—**U**, *W. undulata*; **V**, *W. pusilla*.

solitary, subsessile, with pedicels much elongating in fruit . . . .   7. *W. pusilla*
10. Erect perennials; leaves ± erect and scattered,
lanceolate-subulate, up to 7(–10) mm. long and
1 mm. wide; . . . . . .   10. *W. subaphylla*
Perennials or annuals, if perennial leaves not as
above . . . . ·   . . . 11
11. Erect perennials; leaves numerous and densely
set, linear to narrowly lanceolate, 5–15 mm.
long, 0·5–3 mm. wide; flowers subsessile in ±
contracted often head-like inflorescences . .   11. *W. huttonii*
Perennials or annuals, if perennial leaves and/or
inflorescence not as above 12
12. Gynoecium 3-merous; hypanthium ± 10-nerved;
seeds compressed, ± 2-faced 13
Gynoecium 2–3-merous; hypanthium 5–10-
nerved; seeds trigonous 20
13. Perennials 14
Annuals 17
14. Prostrate to decumbent herbs; leaves up to 6–16
mm. long; inflorescence lax and leafy; corolla
6·5–10 mm. long; style markedly shorter than
the corolla . . . .   14. *W. scottii*
Erect or rarely decumbent herbs; leaves up to 10–
80 mm. long; inflorescence usually ± contracted,
spike- or head-like; corolla 3·5–7·5 mm. long;
style about as long as or longer than the corolla 15
15. Flowers sessile in a strongly contracted inflorescence with a dominant dense terminal head
(fig. 4/A, p. 20); style glabrous below   . .   13. *W. capitata*
Flowers in a lax or contracted, often spike-like
inflorescence, but without a dominant terminal
head; style glabrous or hairy below 16
16. Inflorescence ± contracted, often spike-like; style
hairy below; seeds compressed   . . .   12. *W. napiformis*
Inflorescence lax; style glabrous below; seeds ±
trigonous . . . .   18. *W. abyssinica*
(in part)
17. Flowers sessile in a strongly contracted inflorescence with a dominant dense terminal head
(fig. 4/A, p. 20) . . . . . .   13. *W. capitata*
Flowers in ± lax inflorescences, rarely subsessile in
axillary and terminal clusters   18
18. Leaves linear to ovate; inflorescence lax or with
flowers in axillary and terminal ± loose head-
like clusters; flowers subsessile or on pedicels up
to 20 mm. long; corolla 3·2–7 mm. long; hypanthium ± hairy or glabrous; style glabrous or
hairy below . . . .   15. *W. pulchella*
Leaves lanceolate to ovate, never linear; inflorescence lax with pedicels up to 20(–35) mm. long;
corolla 1·5–3·5 mm. long; hypanthium hirsute;
style glabrous below 19
19. Corolla blue to white; inflorescence usually ±
pyramidal with the main axis distinct nearly to
the top; stem and pedicels hirsute with mixed

hairs of different length; leaves lanceolate to
narrowly ovate . . . . . . 16. *W. erecta*

Corolla yellow; inflorescence spreading, with main
axis not particularly distinct; stem and pedicels
hirsute with long hairs of almost uniform
length; leaves narrowly ovate to ovate . . 17. *W. flexuosa*

20. Perennials; gynoecium 3-merous 21
    Annuals; gynoecium 2–3-merous 23

21. Flowers in usually lax but sometimes ± contracted
    inflorescences; hypanthium ± 10-nerved (fig.
    2/G). . . . . 18. *W. abyssinica*
    Flowers in dense spherical or elongated terminal
    heads; hypanthium 5-nerved 22

22. Erect herbs; leaves 30–60 mm. long, 3–12(–17)
    mm. wide; upper leaves large, forming an
    involucre surrounding the inflorescence . . 19. *W. collomioides*
    Prostrate herbs; leaves 5–17 mm. long, 1–5 mm.
    wide; leaves diminishing in size upwards, not
    forming an involucre . . . . . 20. *W. polycephala*

23. Gynoecium 3-merous 24
    Gynoecium 2-merous 26

24. Flowers in dense spherical or elongated, terminal
    heads, surrounded by an involucre formed by
    the large upper leaves . . . . . 19. *W. collomioides*
    Flowers in lax inflorescences, without an involucre . . . 25

25. Hypanthium 10-nerved, glabrous . . . 18. *W. abyssinica*
    (subsp.
    parvipetala)

    Hypanthium 5-nerved, ± hirsute . . . 21. *W. hirsuta*

26. Slender ± straggling and trailing herbs, glabrous;
    leaves up to 12 mm. long, ± 1 mm. wide. . 25. *W. paludicola*
    Erect herbs, ± hirsute at least below; leaves larger . . . 27

27. Leaves tending to be verticillate at the middle of
    the stem, lanceolate to narrowly ovate; pedicels
    and hypanthia glabrous; corolla up to 2 mm.
    long . . . . 22. *W. hookeri*
    Leaves not verticillate at the middle of the stem,
    linear to lanceolate; pedicels and hypanthia
    glabrous or hairy; corolla usually more than
    2 mm. long 28

28. Leaves usually ± obtuse, mucronulate, up to 18–
    57 mm. long, 5–16 mm. wide, markedly undu-
    late-crenate at the margin . . . . 24. *W. perrottetii*
    Leaves acute or subacute, up to 7–30 mm. long,
    1–5 mm. wide (in East Africa), usually with a
    denticulate, flat or slightly undulate margin . 23. *W. ramosissima*

1. **W. undulata** (*L.f.*) *A. DC.*, Monogr. Camp.: 148 (1830); Thulin in
Symb. Bot. Upsal. 21(1): 76, fig. 6/A, B, 9/A, 11/A, B, 14/A, N, U, 15/B
(1975). Type: South Africa, Cape Province, *Thunberg* in *Herb. Linnaeus*
221 : 28 (LINN, lecto.)

Perennial or annual erect herb, 20–90 cm. tall. Stems few to many, ±
hirsute towards the base or glabrous. Leaves scattered or somewhat crowded
towards the base, sessile, lanceolate to linear, up to 10–70 mm. long, 1·5–
10 mm. wide, acute to subacute with cuneate to truncate base, ± hirsute or

glabrous; margin cartilaginous, ± undulate-crenate; midvein protruding beneath, lateral veins faint. Inflorescence lax; pedicels up to 10–25 mm. long, glabrous; bracts ± ciliate to dentate or glabrous and entire. Hypanthium obconical to hemispherical, ± 10-nerved, glabrous. Calyx-lobes 1·5–6 mm. long, ± ciliate to dentate or glabrous and entire. Corolla campanulate, (5–)8–16 mm. long, blue, white, yellow or various intermediate colours, deeply 5-lobed, puberulous inside near the base, glabrous outside; tube 1·5–5 mm. long. Stamens with filament-bases broadly dilated to almost cross-shaped, ciliate; anthers ± 2·5–5 mm. long. Ovary 3-locular, subinferior; style somewhat shorter than the corolla, hairy or glabrous below; lobes 3 with 3 glands present between their bases. Capsule 3-locular, 3–12 mm. long, ± 10-nerved; valves 3, 1–3 mm. long. Seeds ± elliptic in outline, ± compressed, 0·35–0·55 mm. long; testa reticulate. Fig. 2/A, N, U, p. 6.

TANZANIA. Masai District: Lolkisale, Mar. 1967, *Beesley* 242 !
DISTR. **T2**; Zambia, Malawi, Mozambique, Rhodesia, Angola, South West Africa, Botswana, Swaziland, Lesotho, South Africa (Transvaal, Natal, Orange Free State, Cape Province), Madagascar
HAB. Upland grassland; 2100 m. (0–2600 m. outside East Africa)

SYN. *Campanula undulata* L.f., Suppl.: 142 (1781)
    Wahlenbergia caledonica Sond. in Fl. Cap. 3: 579 (1865); Brenan in Mem. N.Y. Bot. Gard. 8: 492 (1954). Type: South Africa, Orange Free State, Caledon R., *Burke & Zeyher* 1069 (S, iso. !)

2. **W. virgata** *Engl.*, P.O.A. C: 400 (1895); v. Brehm. in E.J. 53: 121 (1915); T.C.E. Fries in N.B.G.B. 8: 397 (1923); W.F.K.: 86 (1948), excl. fig. 65; F.P.S. 3: 72 (1956); E.P.A.: 1055 (1965); U.K.W.F.: 513, 512 (fig.) (1974); Thulin in Symb. Bot. Upsal. 21(1): 81, fig. 2/A, 6/C, D, 12/O, 16 (1975). Type: Malawi, Mt. Mlanje, *Whyte* (B, syn.†, K, lecto. !, G, Z, isolecto. !)

Perennial ± erect herb, up to 70 cm. tall, from a ± woody taproot. Stems few, glabrous or ± hirsute at least towards the base, furrowed, usually with many long ± erect branches. Leaves few, often scale-like, widely scattered on the stem, lanceolate, 2–7(–22) mm. long, 0·5–2(–6) mm. wide, acute, glabrous or ± hirsute; margin cartilaginous, sparsely denticulate. Inflorescence lax; pedicels 1–5 cm. long, glabrous or glabrescent. Hypanthium usually narrowly obconical, 10-nerved, glabrous. Calyx-lobes 1·5–4 mm. long, entire or almost so, rarely ± ciliate. Corolla 8–10(–13) mm. long, white, ± bluish or yellowish, lobed to about ⅔ of the length, puberulous inside near the base, glabrous outside; tube 2–5 mm. long. Stamens with filament-bases markedly dilated to almost cross-shaped, ciliate; anthers 2·5–4 mm. long. Ovary 2-locular, subinferior; style shorter than the corolla, thickened in the upper part, glabrous below; lobes 2, ± 1·5–2·5 mm. long with 2 or 4 glands present between and, if 4, also below their bases. Capsule 2-locular, usually narrowly obconical, 4–10(–16) mm. long; valves 0·8–1·6 mm. long. Seeds narrowly oblong to elliptic in outline, 0·5–0·9 mm. long; testa reticulate. Fig. 3, p. 10.

UGANDA. Acholi District: Langia Mts., Apr. 1943, *Purseglove* 1389 !; Karamoja District: Mt. Morongole, 11 Nov. 1939, *A. S. Thomas* 3316 ! & Napak Mt., June 1950, *Eggeling* 5895 !
KENYA. Elgon, E. slope, 11 Feb. 1948, *Hedberg* 26 !; Ravine District: Nakuru–Eldoret, approaching Equator, 29 Apr. 1961, *Polhill* 394 !; N. Nyeri District: 5 km. SE. of Nanyuki, 16 June 1943, *Moreau* 14 !
TANZANIA. Mt. Meru, E. slope, 16 Jan. 1970, *Thulin* 316 !; Rungwe Mt., near Mwakaleli, 30 Nov. 1958, *Napper* 1183 !; Songea District: Matengo Hills, Lupembe Hill, 11 Jan. 1956, *Milne-Redhead & Taylor* 8211 !
DISTR. **U1**; **K1**–6; **T2**–4, 6–8; Burundi, Sudan, Ethiopia, Zambia, Malawi, Mozambique, Rhodesia, Swaziland, Lesotho and South Africa (Transvaal, Natal)

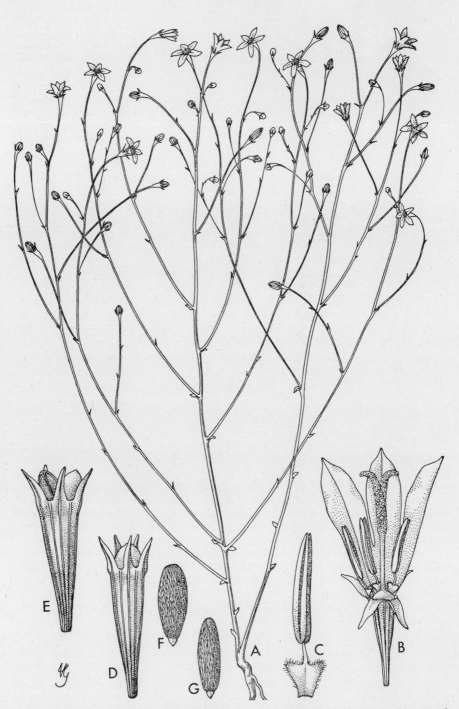

FIG. 3. *WAHLENBERGIA VIRGATA*—**A,** habit, × ⅔; **B,** flower with two stamens and petals removed, × 4; **C,** stamen, × 8; **D,** capsule, × 4; **E,** same, after dehiscence; **F, G,** seed, two views, × 24. A–C, from *A. S. Thomas* 2143; D–G, from *Thulin* 326.

Hab.   Upland grassland, often in patches of open soil, e.g. in eroded places or road-sides; 1100–2700 m.

Syn.   *W. sparticula* Chiov. in Ann. Bot. Roma 10: 389 (1912); E.P.A.: 1055 (1965).
   Type: Ethiopia, Shoa, near Oletta, *Negri* 639 (FI, holo.!)
   W. recurvata v. Brehm. in E.J. 51: 232 (1914) & 53: 119 (1915). Type: Tanzania,
   Mpwapwa District, Nsogiro Mts., *Houy* in *Meyer* 1178 (B, holo. †)
   W. virgata Engl. var. *longisepala* v. Brehm. in E.J. 53: 122 (1915).   Type:
   Tanzania, Uluguru Mts., between Mgeta and Mbakana, *Goetze* 324 (B, holo. †)
   W. virgata Engl. var. *tenuis* v. Brehm. in E.J. 53: 122 (1915).   Type: Tanzania,
   Kilosa District, " Ledingombe ", *Meyer* 1157 (B, holo. †)

3. **W. songeana** *Thulin* in Symb. Bot. Upsal. 21(1): 86, fig. 5/A (1975).
Type: Tanzania, Songea District, Matengo Hills, Lupembe, *Zimmer* 47
(BM, holo.!)

Erect herb $\pm$ 1 m. tall, probably perennial. Stem robust, $\pm$ 4 mm. thick,
sparingly branched, hirsute but less so above. Leaves lanceolate, up to 40
mm. long and 9 mm. wide, hirsute; margin somewhat cartilaginous, dentate; midvein protruding beneath, lateral veins faint. Inflorescence strongly
contracted; flowers subsessile, with ciliate bracts. Hypanthium 10-nerved,
glabrous. Calyx-lobes 2–3·5 mm. long, ciliate. Corolla violet, $\pm$ 6 mm. long,
deeply lobed, hairy inside near the base, glabrous outside; tube $\pm$ 2 mm.
long.   Filament-bases linear, ciliate; anthers $\pm$ 1·2 mm. long. Ovary 3-
locular, subinferior; style shorter than the corolla, hairy below; lobes 3, $\pm$
1·5 mm. long, with 3 tiny glands between their bases. Capsule 3-locular,
10-nerved; valves 3, $\pm$ 1 mm. long. Seeds elliptic in outline, $\pm$ 0·6 mm.
long, almost smooth (only young seeds seen).

Tanzania.   Songea District: Matengo Hills, Lupembe, 30 Aug. 1936, *Zimmer* 47!
Distr. **T8**; known only from the type
Hab.   Upland grassland; 1900 m.

4. **W. lobelioides** (*L.f.*) *A. DC.*, Monogr. Camp.: 157, t. 17 (1830); Thulin
in Symb. Bot. Upsal. 21(1): 89 (1975). Type: Canary Is., *Masson* in *Herb.
Linnaeus* 221: 21 (LINN, lecto.)

Annual $\pm$ erect herb, 10–70 cm. tall. Stem from a taproot, often branched
from the base, glabrous or $\pm$ hirsute below. Leaves alternate or lower ones
sometimes opposite, linear to elliptic or narrowly lanceolate or oblanceolate,
up to 15–85 mm. long, 3–20 mm. wide, acute to subacute, $\pm$ hirsute or
glabrescent, sometimes ciliate; margin thin or $\pm$ cartilaginous, $\pm$ undulate-
dentate; midvein prominent beneath, lateral veins visible or obscure.
Inflorescence lax; pedicels up to 60 mm. long, glabrous. Hypanthium $\pm$
obconical to obovoid, (5–)10-nerved, glabrous. Calyx-lobes 3–5, 0·8–4·8 mm.
long. Corolla 4–7 mm. long, 3–5-lobed, glabrous outside; tube $\pm$ 1·2–3·2 mm.
long. Stamens 3–5; filament-bases narrowly to broadly dilated, $\pm$ ciliate;
anthers $\pm$ 0·6–2·4 mm. long. Ovary 2–3-locular, subinferior; style somewhat
shorter than the corolla, glabrous below; lobes 2–3, with 2–3 minute glands
between their bases.   Capsule 3–12 mm. long, 2–3-locular, (5–)10-nerved,
valves 2–3, 0·6–2·4 mm. long. Seeds elliptic in outline, 0·4–0·6 mm. long;
testa almost smooth.

Syn.   *Campanula lobelioides* L.f., Suppl.: 140 (1781)

   subsp. **nutabunda** (*Guss.*) *Murb.* in Acta Univ. Lund. 33 (12): 115 (1897); Thulin in
Symb. Bot. Upsal. 21(1): 92, fig. 5/D, 9/C, 11/E, F (1975). Type: Italy, Calabria,
Thomas (NAP, holo., G-DC, ? iso.!)

   Stem hirsute in the lower half or at the base only. Leaves oblanceolate to elliptic,
almost glabrous or $\pm$ hirsute, especially beneath and at margins, marginal hairs usually
of smaller size; margin not cartilaginous, $\pm$ flat and dentate. Hypanthium $\pm$ 10-

nerved.   Calyx and corolla (4–)5-lobed.   Stamens (4–)5; filament-bases abruptly but narrowly dilated, ciliate; anthers ± 0·6–1·4 mm. long.   Gynoecium 3-merous.   Capsule ± obconical to obovoid, 3-locular, ± 10-nerved.   Seeds 0·5–0·6 mm. long.

KENYA.   Northern Frontier Province: Mt. Nyiru, W. side, 8 Aug. 1971, *Archer* 697!; Machakos District: Lukenya, 16 June 1968, *Agnew* 10099!
DISTR.   K1, 4; SE. Egypt, NE. Sudan, Ethiopia, Socotra, in the western Mediterranean area from Morocco to Italy; probably identical forms also in South West Africa and South Africa
HAB.   Sandy or rocky places in grassland or bushland; 1950–2250 m.

SYN.   *Campanula nutabunda* Guss. in Tenore, Ad Fl. Neapol. Prodr. App. 5: 8 (1826)
   Wahlenbergia nutabunda (Guss.) A. DC., Monogr. Camp.: 157, t. 17 (1830)
   Laurentia etbaica Schweinf. in Verh. Zool.-Bot. Ges. Wien 18: 683 (1868).   Type: Sudan, Soturba Mts., *Schweinfurth* 1704 (B, holo. †)
   Wahlenbergia etbaica (Schweinf.) Vatke in Linnaea 38: 700 (1874); Hemsl. in F.T.A. 3: 480 (1877); v. Brehm. in E.J. 53: 112 (1915); F.P.S. 3: 72 (1956); E.P.A.: 1055 (1965).   Type as for *Laurentia etbaica*
   [*W. riparia* sensu Balf. f. in Trans. Roy. Soc. Edin. 31: 147 (1888), *non* A. DC.]
   W. nutabunda (Guss.) A. DC. var. *erythraeae* Chiov. in Ann. Bot. Roma 10: 390 (1912).   Type: Ethiopia, Eritrea, Habab, Oazat, *Pappi* 8349 (FI, holo.!)
   W. riparia A. DC. var. *etbaica* v. Brehm. in E.J. 53: 110 (1915).   Types: ?Egypt, Ellagebirge, *Schweinfurth* 1864 & Gebel Cheich Embarak, *Deflers* (?B, syn. †)
   W. sp. A sensu Agnew, U.K.W.F.: 513 (1974)

NOTE.   Subsp. *lobelioides*, with glabrous or glabrescent stem, usually 2-merous gynoecium and 3–4-merous calyx, corolla and androecium, occurs in Madeira, the Canary Is. and the Cape Verde Is. Subsp. *riparia* (A. DC.) Thulin differs by having ± undulate leaves with a cartilaginous margin and occurs in tropical West Africa.

5. **W. silenoides** *A. Rich.*, Tent. Fl. Abyss. 2: 3 (1851); Hemsl. in F.T.A. 3: 478 (1877); Engl. in Abh. Preuss. Akad. Wiss. 1891: 411 (1892); v. Brehm. in E.J. 53: 128 (1915); F.P.S. 3: 71 (1956); E.P.A.: 1055 (1965); U.K.W.F.: 513 (1974); Thulin in Symb. Bot. Upsal. 21(1): 95, fig. 5/F, 6/E, F, 9/B, 12/N, 14/B, I, 15/A (1975).   Type: Ethiopia, Simien, Enchetcab, *Schimper* II. 998 (P, holo.!, BM, BR, FI, G, K, M, S, iso.!)

Perennial herb, up to 60 cm. tall, from a taproot.   Stems many, erect or often prostrate at the base and ascending, few-branched, ± hirsute at the base.   Leaves alternate or lower ones opposite, lanceolate to narrowly ovate, 5–25 mm. long, 1–7 mm. wide, acute, with cuneate to truncate base, ± hirsute; margin cartilaginous, sparsely denticulate, sometimes undulate; midvein prominent beneath, lateral veins obscure.   Inflorescences few-flowered, lax, often reduced to a solitary terminal flower; pedicels up to 30(–60) mm. long, glabrous.   Hypanthium narrowly obconical, ± 10-nerved, glabrous.   Calyx-lobes 1·6–4 mm. long.   Corolla 4–6·5 mm. long, white or tinged with blue or grey-blue, deeply split into ± narrowly elliptic lobes, ± puberulous inside near the base, glabrous outside; tube 1·2–2·2 mm. long.   Filament-bases ± dilated, ciliate; anthers 1·2–2 mm. long.   Ovary 2-locular, subinferior; style shorter than the corolla, slightly thickened in the upper part, eglandular, glabrous or slightly pubescent below; lobes 2, 0·7–1·2 mm. long.   Capsule 2-locular, narrowly obconical, up to 17 mm. long; valves 2, 0·8–3·2 mm. long.   Seeds elliptic in outline, ± compressed, 0·6–0·7 mm. long; testa almost smooth.   Fig. 2/B, I, p. 6.

UGANDA.   Karamoja District: Mt. Moroto, 3 Jan. 1937, *A. S. Thomas* 2147! & Napak Mt., June 1950, *Eggeling* 5902!
KENYA.   Elgon, Koitogoch [Koitagotch], Mar. 1968, *Tweedie* 3534!; Elgeyo District: Cherangani Hills, Kaibwibich, Mar. 1965, *Tweedie* 3020!; Aberdares, Kinangop, Apr. 1938, *Chandler* 2317 in part!
TANZANIA.   Mpanda District: Mahali Mts., Kabesi, 31 Aug. 1958, *Newbould & Jefford* 1969!
DISTR.   U1, 3; K3, 4; T4; Sudan, Ethiopia, Sao Tomé, Fernando Poo, Nigeria, Cameroun, Zaire, Burundi
HAB.   Upland grassland and moor, forest margins; 2200–3350 m.

SYN. *W. polyclada* Hook. f. in J.L.S. 6: 15 (1861), *non* A. DC. (1839), *nom. illegit.*
Type: Fernando Poo, Clarence Peak, *Mann* 600 (K, holo.!, S, iso.!)
W. mannii Vatke in Linnaea 38: 700 (1874); Hemsl. in F.T.A. 3: 478 (1877);
v. Brehm. in E.J. 53: 128 (1915); Hepper in F.W.T.A., ed. 2, 2: 309 (1963).
Type: as for *W. polyclada*
W. silenoides A. Rich. var. *elongata* v. Brehm. in E.J. 53: 128 (1915). Type:
Ethiopia, Simien, *Schimper* 928 (B, syn. †, K, lecto.!, G, P, isosyn.!)

6. **W. krebsii** *Cham.* in Linnaea 8: 195 (1833); Vatke in Linnaea 38: 700
(1874); Thulin in Symb. Bot. Upsal. 21(1): 98 (1975). Type: South Africa,
Cape Province, " Caffraria ", *Krebs* 134 (B, ? holo. †)

Perennial erect or sometimes procumbent herb, 10–50 cm. tall, from a
taproot. Stems few to many, glabrous, ± hirsute or sometimes white
pubescent. Leaves sessile or subpetiolate, ± crowded towards the base or
scattered along the stem, almost linear to oblanceolate or obovate, up to 55
mm. long and 10(–20) mm. wide, acute to subacute with cuneate base,
glabrous or ± hirsute; margin cartilaginous, undulate–dentate; midvein
prominent beneath, lateral veins visible or obscure. Inflorescence lax;
pedicels 2–25 mm. long, glabrous. Hypanthium obconical to hemispherical,
± 10-nerved, glabrous. Calyx-lobes 1·4–4(–4·8) mm. long, entire. Corolla
4·4–11(–16) mm. long, white to blue or violet, often with dark veins, split ±
halfway into lanceolate lobes, puberulous inside near the base, glabrous
outside; tube 2·4–5(–7) mm. long. Stamens with filament-bases narrowly
triangular to broadly dilated, glabrous or ciliate; anthers 0·8–3·5 mm. long.
Ovary subinferior to subsuperior, 3-locular; style shorter than the corolla,
eglandular, thickened in the upper part, glabrous below; lobes 3, 0·4–1·2 mm.
long. Capsule 3-locular, obconical to hemispherical, 3–10 mm. long, ± 10-
nerved; valves 3, up to 3·2 mm. long. Seeds elliptic to broadly elliptic in
outline, ± compressed, 0·5–0·7 mm. long; testa reticulate.

SYN. *W. zeyheri* Buek in Eckl. & Zeyh., Enum. Pl. Afr. Austr. Extratrop. 3: 379
(1837); Sond. in Fl. Cap. 3: 580 (1865); v. Brehm. in E.J. 53: 101 (1915).
Type: South Africa, Cape Province, near Philipstown, *Ecklon & Zeyher* 2374
(G, S, iso.!)

subsp. **arguta** (*Hook. f.*) *Thulin* in Symb. Bot. Upsal. 21(1): 99, fig. 3/B, 5/H, I, 6/I,
J, 9/E, 10/C, D, 12/I, 14/C, M (1975). Type: Fernando Poo, Clarence Peak, *Mann* 601
(K, holo.!, P, S, iso.!)

Differs from subsp. *krebsii*, which occurs in South Africa, in its ± obconical capsules,
which are longer than broad (instead of hemispherical, broader than long), filament-
bases almost linear or slightly broadened and rarely ciliate (instead of broadly dilated
and ciliate), usually smaller and more faintly coloured corolla with narrower lobes, and
in its shorter style-lobes (up to 0·8 mm. versus up to 1·2 mm. long). Fig. 2/C, M, p. 6.

UGANDA. Ruwenzori, 14 Jan. 1967, *Magogo* 29!; Kigezi District: Chahafi–Behungi,
13 Jan. 1933, *C. G. Rogers & Gardner* 364! & Mgahinga, June 1951, *Purseglove* 3695!
KENYA. W. Suk District: Kapenguria, May 1932, *Napier* 1992; Naivasha District:
Kinangop, Sasamua Dam, 2 Sept. 1951, *Verdcourt* 606!; Mt. Kenya, 26 Jan. 1970,
Thulin 330!
TANZANIA. Kilimanjaro, Shira Plateau, 13 Jan. 1970, *Thulin* 312!; Iringa District:
Mufindi, Luisenga, 15 Mar. 1962, *Polhill & Paulo* 1768!; Rungwe Mt., N. slope,
8 Feb. 1961, *Richards* 14320!
DISTR. U2, 3; K2–5; T2, 3, 6, 7; Ethiopia, Fernando Poo, Cameroun, Zaire, Rwanda,
Burundi
HAB. Upland grassland and moor, forest margins; 1500–3500(–4000) m.

SYN. *W. arguta* Hook. f. in J.L S 6: 15 (1861); Hemsl. in F.T.A. 3: 478 (1877); Engl.
in Z.A.E.: 343 (1911); v. Brehm. in E.J. 53: 99 (1915); Hepper in F.W.T.A.,
ed. 2, 2: 309, fig. 273/H–L (1963)
Lightfootia arabidifolia Engl. in E.J. 19, Beibl. 47: 53 (1894). Type: Tanzania,
Kilimanjaro, Oct. 1893, *Volkens* 1116 (B, holo. †, K, lecto.!, BM, BR, G,
iso.!)
Wahlenbergia arabidifolia (Engl.) v. Brehm. in E.J. 53: 99 (1915); T. C. E. Fries
in N.B.G.B. 8: 396 (1923); F.P.N.A. 2: 404, t. 39 (1947); W.F.K.: 86 (1948);

A.V.P.: 186, 333 (1957); E.P.A.: 1054 (1965); U.K.W.F.: 513, 512 (fig.)
(1974)
W. arguta Hook. f. var. *longifusiformis* v. Brehm. in E.J. 53: 100 (1915); R. E.
Fries, Schwed. Rhod.-Kongo-Exped. 2: 315 (1916); F.P.N.A. 2: 404 (1947).
Type: Rwanda, Kissenye, Ninagongo, *Mildbraed* 1403 (B, holo. †)
W. sarmentosa T. C. E. Fries in N.B.G.B. 8: 396, t. 5 (1923). Type: Mt. Kenya,
R. E. & T. C. E. Fries 1178a (UPS, holo. !, BR, K, S, Z, iso. !)

NOTE. The most common East African form of the variable subsp. *arguta* is usually
erect with ± long and narrow leaves tending to be crowded towards the base of the
stem (*W. arabidifolia*).
    W. sarmentosa is a form with straggling or trailing stems and ± evenly scattered
comparatively short and broad leaves found in the bamboo and the *Hagenia-Hypericum*
zones on some of the East African mountains (especially Mt. Kenya and Aberdares).
Intermediates are common, however, and *W. sarmentosa* is probably best regarded
as an ecotype adapted to more mesophytic habitats.
    The type of subsp. *arguta* belongs to a form occurring on Mt. Cameroun and
Fernando Poo. This is characterized by its small leaves and by having capsular
valves about twice as long as the rest of the capsule. It also intergrades with the
" *arabidifolia* " form which is to be found in West Africa too.

7. **W. pusilla** *A. Rich.*, Tent. Fl. Abyss. 2: 2 (1851); Hemsl. in F.T.A. 3:
479 (1877); Engl. in Abh. Preuss. Akad. Wiss. 1891: 412 (1892); v. Brehm.
in E. J. 53: 130 (1915); T. C. E. Fries in N.B.G.B. 8: 397 (1923); A.V.P.:
185, fig. 11 (1957); E.P.A.: 1055 (1965); U.K.W.F.: 511 (1974); Thulin in
Symb. Bot. Upsal. 21(1): 102, fig. 3/D, 6/L, M, 9/F, G, 12/M, 14/L, V (1975).
Type: Ethiopia, Simien, Mt. Buahit above Enchetcab, *Schimper* II. 585
(P, holo. !, BM, BR, FI, G, K, M, S, iso. !)

Dwarf perennial often mat-forming herb with an elongated rhizome that is
much branched in its upper part, each branch ending in a leaf-rosette.
Leaves usually very densely set, sessile or subpetiolate, oblanceolate
to narrowly obovate or spathulate, 5–20(–50) mm. long, 1·5–7(–11) mm. wide,
acute to subacute with attenuate to cuneate base, ± hirsute; margin cartilaginous, ± undulate-dentate; midvein prominent beneath, lateral veins
often obscure. Flowers solitary from leaf-axils, sessile or very shortly
pedicelled; pedicels usually much elongated in fruit and up to 10 cm. long.
Hypanthium hemispherical, 10-nerved, glabrous. Calyx-lobes 1·2–3·6 mm.
long, entire. Corolla 4–6·5 mm. long, white or pale blue with darker veins,
lobed to about ⅓ of the length, puberulous inside near the base; tube 2·5–4
mm. long. Stamens with filament-bases narrowly triangular or almost
linear, glabrous; anthers 0·5–1 mm. long. Ovary subinferior, 3-locular;
style shorter than the corolla, eglandular, glabrous below; lobes 3, ± 0·4 mm.
long. Capsule 3-locular, hemispherical, 10-nerved; valves 3, short. Seeds
elliptic in outline, ± compressed, ± 0·6 mm. long; testa reticulate. Fig.
2/L, V, p. 6.

KENYA. Elgon, E. slope, 23 Feb. 1948, *Hedberg* 140!; Aberdares, E. slope, 8 Oct.
1967, *Hedberg* 4297!; Mt. Kenya, W. slope, 30 Jan. 1922, *R. E. & T. C. E. Fries*
1299!
TANZANIA. Mt. Meru, 16 Jan. 1970, *Thulin* 319!; Kilimanjaro, 10 Dec. 1932, *Geilinger*!
& Shira Plateau, 13 Jan. 1970, *Thulin* 310!
DISTR. **K**3, 4; **T**2; Ethiopia
HAB. Upland grassland and moor, often in open soil in moist places; 2800–4500
(–4900) m.

SYN. *Lobelia kilimandscharica* Engl. in E.J. 19, Beibl. 47: 52 (1894). Type: Tanzania,
Kilimanjaro, S. slope of Mawenzi, *Volkens* 1363 (B, holo. !)

NOTE. *W. pusilla* tends to occur at higher altitudes than the closely related *W. krebsii*
subsp. *arguta*. The altitudinal separation is less pronounced towards the northern
part of the range and in Ethiopia, where *W. pusilla* is to be found down to about
2500 m., they not infrequently grow together. No certain hybrids are known from
nature, however, although partly fertile hybrids have been artificially produced by
experimental cultivation.

8. **W. capillacea** (*L.f.*) *A. DC.*, Monogr. Camp.: 156 (1830); v. Brehm. in E. J. 53 : 94 (1915); Thulin in Symb. Bot. Upsal. 21(1): 111 (1975). Type: South Africa, " e Cap b. Spei ", *Thunberg* (UPS-THUNB, lecto. !)

Perennial erect herb, up to 50 cm. tall, from a thick taproot; stems several, glabrous or ± puberulous; old stems with persistent leaf-bases. Leaves sessile, usually many and densely set, often fascicled (in southern Africa), linear or rarely narrowly oblanceolate, 5–30 mm. long, 0·2–1·2(–3) mm. wide, margin not cartilaginous, flat or ± involute, entire or lower leaves sparsely dentate (in southern Africa rarely narrowly lobed or incised). Inflorescence lax; pedicels 3–25 mm. long, glabrous or puberulous, or rarely flowers subsessile. Hypanthium 10-nerved, glabrous or rarely ± puberulous. Calyx-lobes (1·2–)1·5–4(–6) mm. long, entire, glabrous or ± puberulous. Corolla 5–13 mm. long, blue, lobed to ⅓ or ¼ of the length; tube 3·5–9 mm. long, puberulous inside near the base, sometimes the inside ± densely set with long hairs (in southern Africa). Stamens with filament-bases narrowly dilated, ciliate; anthers 1·5–3·5 mm. long. Ovary 3-locular, semi-inferior; style shorter than the corolla, eglandular, not markedly thickened in the upper part, glabrous or hairy below; lobes 3, ± 0·8 mm. long. Capsule 3-locular, obconical or rarely hemispherical, 10-nerved; valves 2–3 mm. long. Seeds narrowly oblong to broadly elliptic in outline, 0·4–0·9 mm. long, ± wrinkled; testa ± distinctly reticulate.

SYN. *Campanula capillacea* L.f., Suppl.: 139 (1781)

subsp. **tenuior** (*Engl.*) *Thulin* in Symb. Bot. Upsal. 21(1): 113, fig. 1/B, 7/K, L, 9/K, 12/F, 14/D, S, 19/B (1975). Type: Tanzania, Mbeya Mt., *Goetze* 1071 (B, holo. †, BR, lecto. !, BM, iso. !)

Differs from subsp. *capillacea*, which occurs in South Africa (Natal, Cape Province), Swaziland and Lesotho, mainly by the absence of long hairs in the corolla-tube and by its leaves that are never fascicled. Fig. 2/D, S, p. 6.

KENYA. Northern Frontier Province: Mt. Nyiru, June 1936, *Jex-Blake* 5 !; Cherangani Hills, 11 Dec. 1959, *Bogdan* 4987 !; Mt. Kenya, 10 Nov. 1943, *Bally* 3322 !
TANZANIA. Kilimanjaro, 19 Dec. 1935, *Turrall* 77 !; Iringa District: Mt. Image, 28 Feb. 1962, *Polhill & Paulo* 1621 !; Njombe District: Kitulo [Elton] Plateau, May 1953, *Eggeling* 6594 !
DISTR. **K**1–4; **T**2–4, 7; Burundi, Malawi, Mozambique, Rhodesia
HAB. Upland grassland, often in rocky places; 1500–3500 m.

SYN. *W. kilimandscharica* Engl. in Abh. Preuss. Akad. Wiss. 1891: 412 (1892); v. Brehm. in E.J. 53: 78, fig. 2/F–J (1915); W.F.K.: 86, fig. 65 (erroneously named *W. virgata*) (1948); U.K.W.F.: 513, 512 (fig.) (1974), *nom. illegit. confus*. Types: Tanzania, Kilimanjaro, *Meyer* 95 & 150 (B, syn. †)
W. oliveri Schweinf. in Abh. Preuss. Akad. Wiss. 1891: 412 (1892), *nom. nud.*
W. capillacea (L.f.) A. DC. var. *tenuior* Engl. in E.J. 30: 418 (1901); v. Brehm. in E.J. 53: 94, fig. 2/A–E (1915)
W. kilimandscharica Engl. var. *intermedia* v. Brehm. in E.J. 53: 78 (1915). Type: Tanzania, Lushoto District, Mombo, *Grote* 5064 (B, holo. †, EA, lecto. !)
W. aberdarica T. C. E. Fries in N.B.G.B. 8: 395, fig. 5 (1923). Type: Kenya, Aberdares, between Satima and Kinangop, *R. E. & T. C. E. Fries* 2708 (UPS, holo. !)

NOTE. Deviating forms with contracted inflorescences and subsessile flowers occur in southern Tanzania (e.g. *Procter* 1271 and *Richards* 6684).

9. **W. denticulata** (*Burch.*) *A. DC.*, Monogr. Camp.: 152, t. 16 (1830); Thulin in Symb. Bot. Upsal. 21(1): 116, fig. 10/G, 14/H, T (1975). Type: South Africa, Cape Province, Griqualand West, *Burchell* 2000 (K, holo. !, G–DC, iso.!)

Perennial, rarely annual, often subshrubby herb. Stems usually many, ± decumbent and ascending, rarely erect, 0·1–1 m. long, ± hirsute, puberulous or glabrescent; hairs mostly retrorse; old stems with persistent leaf-

bases. Leaves sessile, ± spreading, linear to linear-lanceolate, 3–20(–35) mm.
long, 0·3–2·5 mm. wide, acute to subacute, glabrous or pubescent; margin ±
cartilaginous, usually markedly denticulate. Inflorescence lax or variously
contracted; pedicels up to 20 mm. long, glabrous or shortly pubescent, or
rarely flowers subsessile. Hypanthium often broadly obconical, 10-nerved,
glabrous or pubescent. Calyx-lobes 1·5–5 mm. long, ± recurved and denticulate, glabrous or pubescent. Corolla 5–8(–10) mm. long, blue to white or
yellow, deeply split into linear-lanceolate lobes, puberulous inside near the
base, glabrous or shortly pubescent outside; tube 1–3·2 mm. long. Stamens
with filament-bases ± rhombic, ciliate; anthers 1·5–2·8 mm. long. Ovary
subinferior to semi-superior, 3–10-locular; style usually slightly longer than
the corolla, thickened in the upper part, eglandular, hairy below; lobes
3, ± 0·6–1 mm. long. Capsule 3-locular, 10-nerved. Seeds elliptic in outline,
± compressed or bluntly trigonous, usually very narrowly winged, 0·4–
0·6 mm. long; testa almost smooth. Fig. 2/H, T, p. 6.

KENYA. Machakos District: Lukenya, June 1932, *Mainwaring* in *Napier* 2144! &
   Kapiti Plains, Mwami Hill, 22 June 1957, *Bally* 11527! & base of Mua Hills, 30 May
   1958, *Verdcourt & Napper* 2170!
TANZANIA. Mbulu District: Hanang Mt., 2 Sept. 1932, *B. D. Burtt* 4018!; Kondoa
   District: Kolo, 1 Feb. 1928, *B. D. Burtt* 1291!; Mbeya Airport, 8 Apr. 1956, *Semsei*
   2420!
DISTR. **K**4; **T**2, 4, 5, 7; Zaire, Zambia, Rhodesia, Malawi, Angola, South West Africa,
   South Africa
HAB. Grassland or woodland, roadsides, old cultivations, usually in sandy or rocky
   places; 1050–2600 m.

SYN. *Campanula denticulata* Burch., Trav. 1: 538 (1822)
      Lightfootia denticulata (Burch.) Sond. in Fl. Cap. 3: 559 (1865); Markgraf in
         E.J. 75: 209 (1950); Adamson in Journ. S. Afr. Bot. 21: 170 (1955); Roessler
         in Prod. Fl. SW. Afr. 136: 2 (1966); U.K.W.F.: 511 (1974)
      L. tenuifolia A. DC. in Ann. Sci. Nat., sér. 5, 6: 327 (1866); Hemsl. in F.T.A. 3:
         475 (1877); Hiern, Cat. Afr. Pl. Welw. 1: 629 (1898). Type: Angola, Huila,
         Welwitsch 1157 (G, lecto.!, BM, BR, C, K, LISU, P, isolecto.!)
      Wahlenbergia spinulosa Engl. in E.J. 10: 271 (1888), *non* A. DC. (1830), *nom.
         illegit.* Type: South West Africa, Hereroland, Okahandja, *Marloth* 1337 (B,
         holo. †, K, lecto.!)
      Lightfootia goetzeana Engl. in E.J. 30: 419 (1901). Type: Tanzania, Njombe
         District, Kinjika Diude Mt., *Goetze* 940 (B, holo. †, BR, lecto.!, BM, G, P, iso.!)
      L. laricifolia Engl. & Gilg in Kun.-Samb.-Exped.: 397 (1903). Type: Angola,
         Bié, Lazingua R., *Baum* 837 (B, holo. †, Z, lecto.!, G, iso.!)
      L. denticulata (Burch.) Sond. var. *podanthoides* Markgraf in N.B.G.B. 15: 466
         (1941) & in E.J. 75: 209 (1950). Type: South West Africa, Okavango, Kara-
         kuwisa, *Dinter* 7325 (B, holo. †, M, lecto.!)

NOTE. A very variable species as a whole, but fairly uniform in East Africa. The
   inflorescences are usually lax, but in southern Tanzania there is a form with markedly
   contracted, rounded or oblong inflorescences (*Lightfootia goetzeana*).

10. **W. subaphylla** (*Bak.*) *Thulin* in Symb. Bot. Upsal. 21(1): 124 (1975).
Type: Central Madagascar, *Baron* 2146 (K, holo.!, P, iso.!)

Perennial erect herb, 20–50 cm. tall, from a pale rootstock. Stems few to
many, usually sparsely branched, striate, glabrous or puberulous (usually at
the base only), grey-green. Leaves sessile, ± erect, lanceolate-subulate,
1·5–7(–10) mm. long, 0·5–1 mm. wide; margin ± cartilaginous, entire or with
a few teeth. Inflorescence ± lax or strongly contracted; pedicels up to
5(–8) mm. long, glabrous or puberulous. Hypanthium obconical, ± 10-nerved,
glabrous or puberulous. Calyx-lobes narrowly triangular, 1·2–2·8 mm. long,
acute, glabrous, margin sometimes with a few reflexed teeth. Corolla 5·5–9 mm.
long, white or bluish, deeply split into linear lobes, puberulous inside near
the base, glabrous outside; tube 0·1–0·2 mm. long. Stamens with filament-

bases ± angularly obovate, ciliate-pubescent; anthers 1·6–3·2 mm. long. Ovary 3-locular, semi-inferior; style about as long as the corolla or shorter, eglandular, slightly thickened in the upper part, glabrous below; lobes 3, 1–1·2 mm. long. Capsule 3-locular; valves 3, 1·5–2·5 mm. long. Seeds elliptic in outline, obscurely trigonous, sometimes slightly winged, 0·6–0·8 mm. long; testa almost smooth.

SYN. *Lightfootia subaphylla* Bak. in J.L.S. 20: 193 (1883)

subsp. **thesioides** *Thulin* in Symb. Bot. Upsal. 21(1): 126, fig. 8/E, F, 10/E, F, 14/E, 21 (1975). Type: Tanzania, Rungwe District, Kyimbila, *Stolz* 110 (UPS, holo.!, B, BM, C, G, K, P, S, Z, iso.!)

Stems glabrous or puberulous only at the base, rarely the whole plant puberulous. Leaves many, up to 7(–10) mm. long, often overlapping. Inflorescence lax, usually much branched and many-flowered. Seeds rather bluntly and irregularly trigonous. Fig. 2/E, p. 6.

TANZANIA. Ufipa District: Kito Hill, 20 Nov. 1958, *Napper* 965A!; Njombe District: 30 km. S. of Mlangali, 26 Sept. 1970, *Thulin & Mhoro* 1199!; Songea District: Matengo Hills, Lupembe Hill, 10 Jan. 1956, *Milne-Redhead & Taylor* 8094!
DISTR. T4, 7, 8; Zaire, Malawi, Zambia
HAB. Upland grassland, often appearing after burning; 1350–2700 m.

NOTE. Subsp. *subaphylla*, with fewer, widely scattered leaves up to 3·5(–5) mm. long, occurs in mountains of central Madagascar. Subsp. *scoparia* (Wild) Thulin has weak, little branched stems and strongly contracted inflorescences with usually less than 10 flowers. It is restricted to some mountains on the border between Rhodesia and Mozambique.

11. **W. huttonii** (*Sond.*) *Thulin* in Symb. Bot. Upsal. 21(1): 129, fig. 1/C, 22/B (1975). Type: South Africa, Cape Province, Caffraria, Kreylis County and Katberg, *Bowker* (S, holo.!)

Perennial erect herb or subshrub, 10–35(–50) cm. tall, from a woody root-stock. Stems several to many, puberulous, hirsute or glabrous, often un-branched; old stems with persistent leaf-bases. Leaves numerous and densely set, sessile, usually spreading, linear to narrowly lanceolate, 5–15 mm. long, 0·5–3 mm. wide (in South Africa up to 25 × 8 mm.), acute, glabrous, ciliate or hirsute beneath; margin not or slightly cartilaginous, entire. Flowers rather few in ± contracted often head-like inflorescences or solitary, subsessile. Hypanthium obconical, obscurely 10-nerved, glabrous. Calyx-lobes 2·5–4 mm. long, glabrous or ciliate. Corolla 6–10 mm. long, blue or mauve, deeply split into linear lobes, puberulous inside near the base; tube less than 1 mm. long. Stamens with filament-bases narrowly angularly obovate, ciliate; anthers 2–2·5 mm. long. Ovary 3-locular, semi-inferior; style about as long as the corolla or somewhat shorter, eglandular, glabrous below; lobes 3, ± 0·6–0·8 mm. long. Capsule 3-locular, ± 10-nerved; valves 3, ± 2 mm. long. Seeds elliptic in outline, ± compressed, ± 0·6 mm. long; testa almost smooth.

TANZANIA. Morogoro District: Msumbisi [Msumbi], May 1954, *Semsei* 1715!; Mbeya Mt., Apr. 1959, *Procter* 1204!; Njombe District: Mdapo, Mar. 1954, *Semsei* 1692!
DISTR. T6, 7; Malawi, South Africa (Transvaal, Natal, Orange Free State, Cape Province), Swaziland
HAB. Upland grassland, often in rocky places; 2250–2750 m.

SYN. *Lightfootia huttonii* Sond. in Fl. Cap. 3: 556 (1865); Adamson in Journ. S. Afr. Bot. 21: 207 (1955)
L. lycopodioides Mildbr. in N.B.G.B. 11: 686 (1932), *non* A. DC. (1830), *nom. illegit.* Type: Tanzania, Njombe District, Mbejera, Hagafiro, *Schlieben* 931 (B, holo.!, BM, BR, G, M, P, S, Z, iso.!)

NOTE. The plants from Tanzania and Malawi are generally more narrow-leaved than the South African material. Otherwise the species is fairly uniform, despite the large gap in the distribution.

12. **W. napiformis** (*A.DC.*) *Thulin* in Symb. Bot. Upsal. 21(1): 133, fig. 5/J, 8/G, H, 10/H, I, 12/J, 14/R, 22/C (1975). Type: Angola, Cuanza Norte, Pungo Andongo, Cuanza R., *Welwitsch* 1150 (G, lecto.!, BM, K, LISU, P, isolecto.!)

Perennial or ? biennial herb from a taproot. Stems usually few, erect, rarely ± decumbent or straggling, up to 0·2–1 m. tall, glabrous, hirsute or puberulous. Leaves linear to lanceolate or very narrowly elliptic, up to 10–80 mm. long, 0·8–15 mm. wide, acute, glabrous or ± pubescent; margin cartilaginous, slightly revolute, ± undulate-dentate; midvein prominent beneath, lateral veins obscure or visible. Inflorescence leafy, ± spike-like, sometimes very dense or sometimes with flowers ± loosely clustered in the upper leaf axils; flowers subsessile or on ± hirsute or glabrous up to 5(–23) mm. long pedicels. Hypanthium 10-nerved, ± hirsute or glabrous. Calyx-lobes 1·2–3·5 mm. long, ± hirsute or glabrous, usually with ± cartilaginous and sometimes shallowly denticulate margins. Corolla 3·5–7·5 mm. long, blue, purplish, yellowish or whitish, split almost to the base into linear lobes, puberulous inside near the base, ± hirsute or glabrous outside. Stamens with filament-bases ± broadly dilated, ciliate; anthers 1·6–2·8 mm. long. Ovary 3-locular, subinferior to semi-superior; style about as long as the corolla, eglandular, hairy or rarely glabrous below; lobes 3, 0·4–0·8 mm. long. Capsule 3-locular, 2·8–5·5 mm. long, 10-nerved; valves 3, 1·2–3 mm. long. Seeds elliptic-oblong in outline, ± compressed and 2-faced, 0·4–0·7 mm. long, almost smooth. Fig. 2/R, p. 6.

UGANDA. W. Nile District: Koboko, Mar. 1941, *Purseglove* 1114!; Masaka District: Katera–Kyebe [Kiebbe] road, 1·5 km. from Katera, 1 Oct. 1953, *Drummond & Hemsley* 4512! & NW. side of Lake Nabugabo, 9 Oct. 1953, *Drummond & Hemsley* 4678!

KENYA. W. Suk District: Kapenguria, Mar. 1935, *Thorold* 3209!; Machakos District: Kilima Kiu, 14 July 1957, *Bally* 11570!; Kwale District: Shimba Hills, Mwele Mdogo Forest, 4 Feb. 1953, *Drummond & Hemsley* 1118!

TANZANIA. Bukoba District: Maruku road, Aug. 1931, *Haarer* 2090!; Tanga District: Kange, 7 Aug. 1958, *Faulkner* 2169!; Songea District: Matagoro Hills, 22 Feb. 1956, *Milne-Redhead & Taylor* 8872!

DISTR. U1–4; K2–7; T1–5, 7, 8; Ethiopia, Sudan, Central African Republic, Congo (Brazzaville), Zaire, Rwanda, Burundi, Zambia, Malawi, Mozambique, Rhodesia, Angola, South West Africa, Botswana

HAB. Deciduous woodland or grassland, old cultivations, roadsides, usually in sandy or rocky soils; 0–2250 m.

SYN. *Lightfootia napiformis* A. DC. in Ann. Sci. Nat., sér. 5, 6: 328 (1866); Hemsl. in F.T.A. 3: 475 (1877); Hiern, Cat. Afr. Pl. Welw. 1: 629 (1898)
   L. marginata A. DC. in Ann. Sci. Nat., sér. 5, 6: 326 (1866); Hemsl. in F.T.A. 3: 474 (1877); Hiern, Cat. Afr. Pl. Welw. 1: 629 (1898); Engl. & Gilg in Kun.-Samb.-Exped.: 396 (1903); Lambinon & Duvign. in B.S.B.B. 93: 45 (1961), pro parte. Type: Angola, Huila, between Catumba and Ohai, *Welwitsch* 1156 bis (G, lecto.!, BM, LISU, isolecto.!)
   L. glomerata Engl. in E.J. 19, Beibl. 47: 52 (1894); P.O.A. C: 400 (1895); E.P.A.: 1055 (1965), excl. syn.; U.K.W.F.: 511 (1974). Type: Tanzania, Tanga District, Duga, *Holst* 3182 (B, holo. †, G, lecto.!, K, M, P, Z, iso.!)
   L. abyssinica A. Rich. var. *glaberrima* Engl., P.O.A. C: 400 (1895). Type: Sudan, Equatoria, Niamniam, Nabambissoo R., *Schweinfurth* 2995 (B, syn. †, K, lecto.!)
   L. abyssinica A. Rich. var. *cinerea* Engl. & Gilg in Kun.-Samb.-Exped.: 397 (1903). Type: Angola, Bié, Longa R. at Jonkoa, *Baum* 553a (B, holo. †, G, lecto.!, BM, K, Z, iso.!)
   L. kagerensis S. Moore in J.L.S. 37: 176 (1905). Type: Uganda, Masaka District, near mouth of Kagera R., *Bagshawe* 566 (BM, holo.!)
   L. campestris Engl. in Z.A.E.: 343 (1911). Type: Rwanda, Katojo, between Kifumbiro and Mischenye, *Mildbraed* 266 (B, holo. †)
   L. graminicola M. B. Scott in K.B. 1915: 45 (1915). Type: Angola, Huila, Humpata, *Pearson* 2776 (K, holo.!)
   L. marginata A. DC. var. *lucens* Lambinon in B.S.B.B. 93: 46 (1961). Type: Rhodesia, vicinity of Umvukwe Mts., near Darwindale, *Rodin* 4331 (K, holo.!, S, iso.!)

NOTE.   W. *napiformis*, over its wide geographical and habitat range, shows much
variation especially in degree of contraction of the inflorescence, indumentum and
hairiness of the style.   Two principal form series may roughly be distinguished.
(1) The forms in eastern Tanzania and east of the Great Rift Valley in Kenya have
± dense spike-like inflorescences, hairy styles, and are usually hirsute all over (*Lightfootia glomerata*).   Similar forms, although sometimes with distinctly pedicelled
flowers, occur in all eastern Africa from Ethiopia to Rhodesia and Mozambique and
also across southern tropical Africa to Angola (*L. marginata*, *L. napiformis*, *L. abyssinica* var. *cinerea*, etc.).   (2) In the western parts of Ethiopia, Kenya and Tanzania,
in Uganda and northwestern Zambia and further westwards to Zaire, Congo (Brazzaville) and the Central African Republic there are forms with generally more lax and
glabrous inflorescences and often glabrous styles (*Lightfootia abyssinica* var. *glaberrima*,
L. campestris, *L. kagerensis*).   In Kenya these two form series are rather distinct,
but in Tanzania, for example, they ± imperceptibly intergrade and therefore in my
opinion do not merit taxonomic distinction.
    Especially the western forms of *W. napiformis* with ± lax inflorescences may be
difficult to distinguish from *W. abyssinica*.   The only reliable character in some cases
is the compressed, 2-faced (not trigonous) seeds of *W. napiformis*.

13. **W. capitata** (*Bak.*) *Thulin* in Symb. Bot. Upsal. 21(1) : 139, fig. 12/G,
23 (1975).   Type : Malawi, Nyika Plateau, *Whyte* (K, holo. !, G, P, iso. !)

Annual or usually perennial (although probably often short-lived) ±
erect herb, up to 1 m. tall, from a taproot. Stem with long branches from
near the base or unbranched, ± ribbed, hirsute. Leaves sessile, linear or
narrowly lanceolate to lanceolate or elliptic, up to 15–60(–80) mm. long,
1·5–10(–15) mm. wide, acute, with truncate to cuneate base, ± hirsute,
especially on veins beneath or glabrescent ; margin cartilaginous, dentate,
often ± undulate ; midvein prominent beneath, lateral veins less so, but
usually clearly visible. Inflorescence leafy, head-like, sometimes only a
terminal head present, but usually also a number of smaller lateral sessile or
pedunculate ones ; flowers subsessile or rarely some flowers on pedicels up to
1·5 mm. long. Hypanthium obconical, ± 10-nerved, ± hirsute or rarely
glabrescent. Calyx-lobes 2·4–4 mm. long, ciliate-pubescent, rarely glabrescent ; margins cartilaginous. Corolla 5–6·5 mm. long, blue, white or mauve,
split almost to the base into linear lobes, ± hirsute outside on the midvein.
Stamens with filament-bases broadly dilated, sometimes almost cross-
shaped, ciliate-pubescent ; anthers 1·2–2·4 mm. long. Ovary 3-locular, semi-
inferior ; style about as long as or somewhat longer than the corolla, eglandular, slightly thickened in the upper part, glabrous below ; lobes 3, 0·5–0·8 mm.
long. Capsule 3-locular, ± 10-nerved ; valves 3, 1·5–2·5 mm. long. Seeds
elliptic-oblong in outline, compressed, 0·5–0·8 mm. long ; testa almost
smooth. Fig. 4, p. 20.

UGANDA.   Bunyoro District : Bugoma, 19 Dec. 1906, *Bagshawe* 1380 ! & Hoima–Fort
    Portal road, near Kafu R., 25 Sept. 1960, *Lind* 2752 ! ; Kigezi District : Maziba–
    Rwanda border, Dec. 1944, *Purseglove* 1605 !
TANZANIA.   Mpanda District : Mahali Mt., Utahya, 6 Sept. 1958, *Jefford & Newbould*
    2387 ! ; Iringa District : Mufindi, Ngowasi Lake, 23 Mar. 1962, *Polhill & Paulo* 1849 ! ;
    Songea District : Matengo Hills, Miyau, 23 May 1956, *Milne-Redhead & Taylor* 10291 !
DISTR. U2 ; T1, 4, 7, 8 ; Zaire, Rwanda, Burundi, Malawi, Zambia, Rhodesia, Mozam-
    bique
HAB.   Deciduous woodland, bushland and upland grassland, old cultivations, usually
    on sandy soils ; 850–2400 m.

SYN.   *Lightfootia glomerata* Engl. var. *subspicata* Engl. in P.O.A. C : 400 (1895).   Type :
        Malawi, without precise locality, *Buchanan* 40 (K, lecto. !, BM, G, iso. !)
    L. capitata Bak. in K.B. 1898 : 158 (1898)
    L. bequaertii De Wild. & Ledoux in Contrib. Fl. Katanga, Suppl. 3 : 145 (1930).
        Type : Zaire, Katanga, Lubumbashi, *Bequaert* 349 (BR, holo. !)
    [*L. glomerata* sensu Brenan in Mem. N.Y. Bot. Gard. 8 : 491 (1954) ; Lambinon
        & Duvign. in B.S.B.B. 93 : 47 (1961) ; E.P.A. : 1055 (1965), pro parte, *non*
        Engl.]

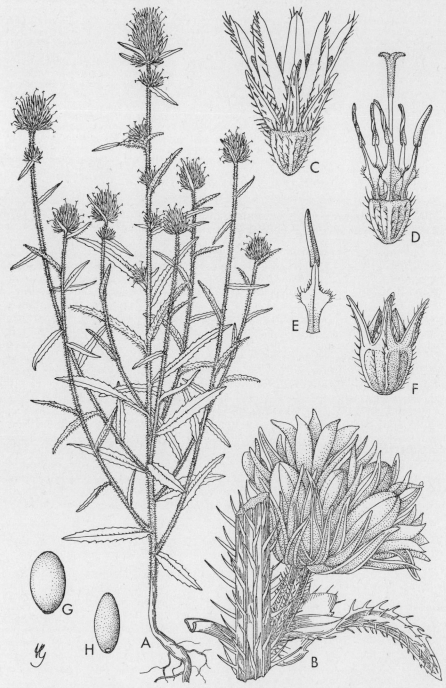

Fig. 4. *WAHLENBERGIA CAPITATA*—**A,** habit, × ⅔; **B,** detail of leaf and stem with young axillary inflorescence, × 6; **C,** young flower (style not shown), × 6; **D,** flower with calyx-lobes and corolla removed, × 6; **E,** stamen, × 8; **F,** capsule, × 6; **G, H,** seed, two views, × 30. A–E, from *Milne-Redhead & Taylo* 10160; F–H, from *Thulin & Mhoro* 1174.

L. glomerata Engl. var. *capitata* (Bak.) Lambinon in B.S.B.B. 93: 47 (1961), pro parte, excl. syn. *L. polycephala*

?*L. elegans* Gilli in Ann. Nat. Mus. Wien 77: 56 (1973).   Type: Tanzania, Njombe District, Lupingu, *Gilli* 555 (W, holo.!)

NOTE.   *Lightfootia elegans* is an annual unusually slender form with long linear leaves. It differs most markedly from normal *W. capitata* by being almost entirely glabrous with a few hairs at the base of the stem only.   At present I regard it as a form of *W. capitata*, but additional material is desirable.

14. **W. scottii** *Thulin* in Symb. Bot. Upsal. 21(1): 145, fig. 2/C, 11/J, K, 24 (1975).   Type: Kenya, NW. of Mt. Kenya, *Battiscombe* 735 (K, holo.!, EA, iso.!)

Perennial decumbent herb with a ± woody taproot.   Stems many, up to 35 cm. long, hirsute or glabrous.   Leaves sessile, lanceolate, 6–16 mm. long, 2–6 mm. wide, acute to obtuse, with truncate base, hirsute beneath, glabrescent above or entirely glabrous; margin cartilaginous, ± undulate and sparsely denticulate.   Inflorescence lax, ± leafy, sometimes reduced to a single terminal flower; pedicels up to 22 mm. long, hirsute or glabrous. Hypanthium obconical, ± 10-nerved, hirsute or glabrous.   Calyx-lobes 2·5–7 mm. long, glabrous to sparsely hirsute and denticulate.   Corolla 6·5–10 mm. long, blue, split almost to the base into linear lobes, puberulous inside towards the base, usually ± hirsute outside on the midvein.   Filament-bases dilated, ciliate; anthers ± 2 mm. long.   Ovary 3-locular, semi-inferior; style shorter than or rarely as long as the corolla, glabrous below; lobes 3, ± 1 mm. long.   Capsule 3-locular, ± 10-nerved; valves 3, 1·5–2 mm. long. Seeds elliptic-oblong in outline, slightly compressed, 0·5–0·6 mm. long; testa almost smooth.

KENYA.   Northern Frontier Province: Lorogi, 23 Oct. 1935, *Leakey* 59!; Elgeyo District: Cherangani Hills, Kaibwibich, Aug. 1968, *Thulin & Tidigs* 131!; Meru District: 16 km. SSW. of Meru, 23 Feb. 1957, *Bogdan* 4451!
TANZANIA.   Masai District: Ngorongoro Crater, 24 July 1957, *Bally* 11595!; Mbulu District: Mt. Hanang, 26 Dec. 1929, *B. D. Burtt* 2267! & Nou Forest Reserve, 24 Sept. 1959, *Kerfoot* 1347!
DISTR.   K1, 3–6; T2; not known elsewhere
HAB.   Upland grassland; 1500–3000 m.

SYN.   *Lightfootia cartilaginea* M. B. Scott in K.B. 1915: 45 (1915); U.K.W.F.: 511, 512 (fig.) (1974), *non Wahlenbergia cartilaginea* Hook. f.

NOTE.   Some plants from the Masai District of Kenya (*Bally* 4186, *Archer* 679) deviate from the normal form by being entirely glabrous on stems, leaves, hypanthium and outside of the corolla.

15. **W. pulchella** *Thulin* in Symb. Bot. Upsal. 21(1): 147 (1975).   Type: Burundi, Luvironza, Gihanga Hill, *Michel* 5532 (BR, holo.!)

Annual ± erect often ± spreading or straggling herb, 5–40 cm. tall.   Stem with few to many usually long branches, rarely unbranched, ± hirsute along its entire length or glabrous in the upper part, rarely all glabrous.   Leaves linear to ovate (basal leaves usually short and broad), up to 7–40 mm. long, 1·5–10 mm. wide, acute to subacute, with cuneate to truncate base, ± hirsute, especially beneath; margin cartilaginous, ± dentate; midvein prominent beneath, lateral veins ± obscure.   Inflorescence lax or with flowers in ± dense axillary and terminal clusters; flowers subsessile or on glabrous or hairy pedicels up to 20 mm. long.   Hypanthium ± obconical, ± 10-nerved, ± hairy or glabrous.   Calyx-lobes 0·8–3·2 mm. long, ± ciliate-pubescent or glabrous; margin cartilaginous.   Corolla 3–6·5 mm. long (outside East Africa 1·5–7 mm.), blue, white or yellow, split almost to the base into linear lobes, ± hairy or glabrous outside.   Filament-bases ± broadly dilated, ciliate; anthers 0·5–1·6 mm. long. Ovary 3-locular, semi-inferior; style about as long as or somewhat shorter than the corolla, eglandular, slightly thickened in the upper part, glabrous or rarely hairy below; lobes 3, 0·3–0·6 mm. long. Capsule 3-locular, ± 10-nerved; valves 3, 0·8–2·5 mm. long. Seeds elliptic-oblong in outline, compressed, 0·4–0·7 mm. long; testa almost smooth.

Key to infraspecific variants

Stem ± hirsute below, but glabrous in the upper part; flowers with distinct pedicels up to 3–20 mm. long; leaves (excluding basal ones) linear to lanceolate; corolla usually glabrous outside . . . . .   subsp. **paradoxa**

Stem ± hirsute along its entire length; flowers sub-sessile or pedicelled; leaves never linear; corolla usually hairy outside:

Flowers with distinct pedicels up to 3–12 mm. long in ± lax, paniculate inflorescences; style hairy or glabrous below. . . . . .   subsp. **pedicellata**

Flowers very shortly (± 0·5 mm.) or occasionally distinctly pedicelled, ± densely grouped in axillary and terminal clusters; style glabrous below   subsp. **mbalensis**

subsp. **paradoxa** *Thulin* in Symb. Bot. Upsal. 21(1): 153, fig. 26/A (1975). Type: Tanzania, Kigoma District, 58 km. S. of Uvinza, *Bullock* 3289 (K, holo.!)

Erect herb, 10–35 cm. tall. Stem ± hirsute below, but glabrous or almost so in the upper part. Leaves lanceolate to linear, up to 11–40 mm. long, 2–6(–8) mm. wide. Inflorescence ± lax; pedicels up to 3–20 mm. long, glabrous. Hypanthium ± hirsute or glabrous. Calyx-lobes 1–2·4 mm. long. Corolla 3·5–6·4 mm. long, blue, glabrous or rarely hairy outside. Anthers 1–1·6 mm. long. Style glabrous or rarely hairy below. Seeds 0·4–0·7 mm. long.

Tanzania. Kigoma District: between Mpanda and Uvinza, N. of Kafulu, July 1951, *Eggeling* 6177!; Mpanda District: near Mpanda, 26 May 1957, *Nye* 199!; Iringa District: Nyangolo scarp, 17 Apr. 1962, *Polhill & Paulo* 2040!
Distr. T4, 5, 7; Zambia
Hab. Sandy grassland, deciduous woodland, eroded places; 1200–1700 m.

Note. There are a number of specimens from central Tanzania ± intermediate between subsp. *paradoxa* and subsp. *pedicellata*, e.g. *Burtt* 3631, *Lindeman* 350A and *Tallantire* 266. These are ± hirsute also in the inflorescence region, have broader leaves than normal in subsp. *paradoxa*, and often hairy styles.

subsp. **pedicellata** *Thulin* in Symb. Bot. Upsal. 21(1): 153, fig. 14/J, 26/C (1975). Type: Burundi, Gitega, Rukongwa, *Van der Ben* 2551 (BR, holo.!, G, K, WAG, Z, iso.!)

More or less erect, often spreading herb, 10–35 cm. tall. Stem hirsute along its entire length. Leaves lanceolate to narrowly ovate or narrowly elliptic, up to 14–30 mm. long, 5–10 mm. wide. Inflorescence ± lax; pedicels up to 3–8 mm. long, hirsute or glabrous. Hypanthium ± hirsute. Calyx-lobes 1·6–2·4 mm. long. Corolla 3·2–5 mm. long, blue or yellow. Style glabrous or hairy below. Seeds ± 0·5–0·7 mm. long. Fig. 2/J, p. 6.

Tanzania. Ngara District: Bugufi, Jan. 1936, *Chambers* K.48! & Mu Rukarazo, 26 Apr. 1960, *Tanner* 4898! & Nyamyaga, 15 July 1960, *Tanner* 5237!
Distr. T1; Rwanda, Burundi
Hab. Deciduous woodland or grassland, old cultivations, often on sandy ground or in rock crevices; 1600–1800 m.

subsp. **mbalensis** *Thulin* in Symb. Bot. Upsal. 21(1): 150, fig. 26/B (1975). Type: Zambia, Mbala District, Kalambo Falls, *Richards* 23219 (K, holo.!, BR, P, iso.!)

Erect herb, 10–40 cm. tall. Stem hirsute along its entire length. Leaves lanceolate to ovate or ± narrowly elliptic, up to 14–32 mm. long, 5–10 mm. wide. Inflorescence with flowers in axillary and terminal clusters; pedicels ± 0·5 mm. long, but occasionally up to 7 mm. long, ± hirsute. Hypanthium ± hirsute. Calyx-lobes up to 2–3·2 mm.

long. Corolla 3·5–5·5 mm. long, blue or white, ± hairy outside. Anthers (0·8–)1·2–1·6 mm. long. Style glabrous below. Seeds 0·5–0·6 mm. long.

TANZANIA. Ufipa District: Kalambo R., near Timber Camp, 16 May 1962, *Richards* 16467!
DISTR. **T4**; Zambia
HAB. Deciduous woodland or bushland, often on sandy soils or among rocks; ± 1200 m. (900–2250 m. in Zambia)

NOTE. A tiny individual from **T7** (*Bjørnstad* 1764 A, Mbeya District, Magangwe) may also belong to subsp. *mbalensis*. The corolla, however, is only ± 2 mm. long and equalling the calyx-lobes and the leaves 7–8 by 2–4 mm. More material is required.
    In addition to the subspecies treated here, subsp. *pulchella* occurs in Burundi, subsp. *laurentii* Thulin in Zaire and subsp. *michelii* Thulin in Rwanda, Burundi and Zaire.

16. **W. erecta** (*Roem. & Schultes*) *Tuyn* in Fl. Males., ser. 1, 6: 113, fig. 1/h-i (1960); E.P.A.: 1054 (1965), quoad syn. pro parte; Thulin in Symb. Bot. Upsal. 21(1): 155, fig. 5/M, 27/A, C, 29/F, G (1975). Type: " India orientali ", *B. Heyne* (B, holo. †, K-WALL, lecto. !)

Annual erect herb, 6–35 cm. tall, with main stem usually distinct nearly to the top of the plant, hirsute with mixed hairs of different length. Branches usually numerous, spreading. Leaves sessile, lanceolate to narrowly ovate or ovate (the lowest only), 8–27 mm. long, 3–8 mm. wide, acute, with cuneate to truncate base, ± hirsute; margin cartilaginous, undulate-dentate; midvein prominent beneath, lateral veins almost invisible. Inflorescence lax, ± pyramidal; pedicels 3–12(–20) mm. long, hirsute with mixed hairs of different length, rarely glabrous. Hypanthium hemispherical, 10-nerved, hirsute. Calyx-lobes 1·5–3 mm. long, usually abruptly broadened at the base, hirsute in lower part only. Corolla 1·5–2·5 mm. long, ± pale blue or white, split almost to the base into linear-lanceolate lobes, ± hirsute outside. Stamens with filament-bases usually narrowly triangular, sometimes slightly broadened, glabrous or sparsely and shortly ciliate; anthers 0·4– 0·7 mm. long. Ovary 3-locular, semi-inferior; style about as long as the corolla, slightly thickened just below the 3 short lobes, eglandular, glabrous below. Capsule 3-locular, 10-nerved; valves 3, 1·2–1·6 mm. long. Seeds elliptic-oblong in outline, compressed, 0·5–0·6 mm. long; testa almost smooth. Fig. 5/F, G, p. 28.

KENYA. W. Suk District: Sekerr Mts., 1 Aug. 1968, *Agnew et al.* 10316!; Fort Hall District: Thika, Blue Posts Hotel, 9 Aug. 1967, *Faden* 67/552!; Nairobi District: Langata, 19 July 1974, *Ngweno* 77!
TANZANIA. Mbeya District: 11 km. W. of Mbeya, 12 May 1956, *Milne-Redhead & Taylor* 10058!; Iringa District: Great North Road, 200 km. S. of Iringa, 28 Mar. 1962, *Polhill & Paulo* 1925!; Songea District: Luhira R. near Mshangano fish ponds, 25 Apr. 1956, *Milne-Redhead & Taylor* 9814!
DISTR. **K2**, 4, 6; **T2**, 3, 5, 7, 8; Sudan, Ethiopia, Nigeria, Angola, Zambia, Malawi, Rhodesia, India, Malaysia
HAB. Grassland or deciduous woodland, waste places, cultivations, often in sandy soil; 800–1500(–2300) m.

SYN. *Dentella erecta* Roem. & Schultes, Syst. Veg. 5: 25 (1819); Roth, Nov. Pl. Sp.: 140 (1821); Cham. & Schlecht. in Linnaea 4: 151 (1829)
    D. perotifolia Roem. & Schultes, Syst. Veg. 5: 25 (1819), *nom. invalid.*, not accepted by the author. Type: " India orientale ", *Klein in Herb. Willd.* 4121 (B, holo., UPS, photo.!)
    Wahlenbergia perotifolia Wight & Arn., Prod. Fl. Pen. Ind. Or. 1: 405 (1834); Wight, Ic. Pl. Ind. 3: 4, t. 842 (1843), *nom. illegit. superfl.* Type: as for *W. erecta*
    Cephalostigma schimperi A. Rich., Tent. Fl. Abyss. 2: 2 (1851); C. B. Cl. in Fl. Brit. Ind. 3: 428 (1881); Hiern, Cat. Afr. Pl. Welw. 1: 628 (1898). Type: Ethiopia, Tigre, *Quartin Dillon & Petit* (P, lecto.!)
    Lightfootia arenaria A. DC. in Ann. Sci. Nat., sér. 5, 6: 329 (1866); Hemsl. in F.T.A. 3: 476 (1877). Type: Angola, Huila, Monino, *Welwitsch* 1155 (G, holo.!, BM, BR, C, K, LISU, P, iso.!)

Cephalostigma erectum (Roem. & Schultes) Vatke in Linnaea 38: 699 (1874),
   quoad syn. pro parte; Vatke in Linnaea 40: 201 (1876), pro parte; Engl. in
   Abh. Preuss. Akad. Wiss. 1891: 408 (1892), pro parte; Brenan in Mem. N.Y.
   Bot. Gard. 8: 490 (1954), pro parte
[*C. hirsutum* sensu Hemsl. in F.T.A. 3: 472 (1877), *non* Edgew.]
C. perotifolium (Wight & Arn.) Hutch. & Dalz., F.W.T.A. 2: 191 (1931), *nom.
   illegit.*, quoad specim. cit. pro parte
[*C. perrottetii* sensu Hepper in F.W.T.A., ed. 2, 2: 311 (1963), pro parte, *non*
   A. DC.]
Lightfootia perotifolia (Roem. & Schultes) Agnew, U.K.W.F.: 511 (1974), *nom.
   invalid.*

17. **W. flexuosa** (*Hook. f. & Thomson*) *Thulin* in Symb. Bot. Upsal. 21(1):
158, fig. 2/D, 5/L, 27/B, 29/H, I (1975). Type: India, Bombay, *Dalzell* (K,
holo.!)

Annual erect herb 7–35 cm. tall, diffusely branched with the main stem not
particularly distinct, hirsute with long hairs of almost uniform length.
Leaves sessile, narrowly ovate to ovate or elliptic, 8–30(–38) mm. long,
5–16(–19) mm. wide, acute, with rounded to cuneate base, ± hirsute;
margin cartilaginous, undulate-dentate; midvein prominent beneath as are
the lateral veins at least on larger leaves. Inflorescence lax; pedicels 5–20
(–35) mm. long, hirsute with long hairs of almost uniform length or glabrous.
Hypanthium hemispherical, 10-nerved, hirsute. Calyx-lobes 1·2–3·2 mm.
long, often broadened at the base, ± hirsute. Corolla 1·5–3·5 mm. long,
yellow, split almost to the base into linear-lanceolate lobes, ± hirsute outside.
Stamens with filament-bases distinctly broadened, ciliate; anthers 0·6–
1·6 mm. long. Ovary 3-locular, semi-inferior; style about as long as the
corolla, eglandular, slightly thickened below the 3 short lobes, glabrous
below. Capsule 3-locular, 10-nerved; valves 3, ± 2 mm. long. Seeds elliptic-
oblong in outline, compressed, ± 0·6 mm. long; testa almost smooth. Fig.
5/H, I, p. 28.

Uganda. Karamoja District: Iriri–Napak Mt., Aug. 1963, *J. Wilson* 1399; Toro
   District: near Fort Portal, Nyakasura, 4 Dec. 1934, *G. Taylor* 2306! & near Fort
   Portal, *Liebenberg* 993!
Tanzania. Lushoto District: W. Usambara Mts., Sakare, 24 Sept. 1902, *Engler* 979
   in part!; Songea District: 2·5 km. NE. of Kigonsera, 15 Apr. 1956, *Milne-Redhead &
   Taylor* 9658!
Distr. **U**1, 2; **T**3, 8; Ethiopia, Cameroun, Nigeria, Central African Republic, Zaire,
   Malawi, India
Hab. Grassland or deciduous woodland, old cultivations; 800–1500 m. (400–2100 m.
   outside East Africa)

Syn. *Cephalostigma flexuosum* Hook. f. & Thomson in J.L.S. 2: 9 (1857); C. B. Cl. in
   Fl. Brit. Ind. 3: 428 (1881)
   C. erectum (Roem. & Schultes) Vatke var. *luteum* Chiov. in Ann. Bot. Roma 9:
   79 (1911). Type: Ethiopia, Beghemder, Dembia, near Asoso, *Chiovenda* 2013
   (FI, holo.!)
   [*Lightfootia hirsuta* sensu Hepper in K.B. 15: 61 (1961), pro parte, & in F.W.T.A.,
   ed. 2, 2: 311 (1963), pro parte]

18. **W. abyssinica** (*A. Rich.*) *Thulin* in Symb. Bot. Upsal. 21(1): 160, fig.
8/I, 10/J, 12/A, B, 14/G, P, Q, 22/A (1975). Type: Ethiopia, Tigre, Shire
[Chiré], *Quartin Dillon & Petit* (P, lecto.!)

Perennial or annual herb from a taproot, which becomes ± long and woody
in old perennating specimens. Stems few–many, erect or rarely ± decumbent,
12–90 cm. tall, glabrous, hirsute or puberulous (hairs often ± curled). Leaves
sessile, linear to lanceolate or very narrowly elliptic, up to 8–75 mm. long,
1–12 mm. wide, acute, glabrous or ± pubescent; margin cartilaginous,
usually slightly revolute, ± undulate-dentate; midvein protruding beneath,

lateral veins obscure or prominent. Inflorescence lax, or sometimes ±
contracted, rarely almost spicate; pedicels up to 2–15 mm. long, glabrous
or rarely pubescent. Hypanthium (7–)10-nerved, glabrous or rarely pubescent. Calyx-lobes 0·8–4 mm. long, usually glabrous, sometimes with markedly
cartilaginous margins. Corolla 2–8 mm. long, blue, white, yellowish or sometimes reddish, split almost to the base into linear lobes, puberulous inside
near the base, glabrous outside. Stamens with filament-bases ± broadly
dilated, ciliate; anthers 0·7–2·5 mm. long. Ovary 3-locular, semi-inferior;
style about as long as the corolla, eglandular, glabrous or hairy below; lobes
3, 0·1–0·8 mm. long. Capsule 3-locular, 2·4–6 mm. long, (7–)10-nerved;
valves 3, 0·5–2 mm. long. Seeds elliptic in outline, usually with acute ends,
± trigonous, 0·4–0·7 mm. long; testa almost smooth.

subsp. **abyssinica**

Perennial erect or rarely ± decumbent herb, up to 90 cm. tall. Leaves linear to
lanceolate, up to 8–75 mm. long, 1–12 mm. wide. Calyx-lobes 1–4 mm. long. Corolla
3·5–8 mm. long, blue, white, yellowish or sometimes reddish. Style glabrous or hairy
below, with lobes (0·2–)0·3–0·8 mm. long. Seeds trigonous, sometimes obscurely so,
0·4–0·7 mm. long. Fig. 2/G, P, Q, p. 6.

Kenya. Northern Frontier Province: Mathews Peak [Ol Doinyo Lengio], 20 Dec. 1958,
Newbould 3276!; Naivasha District: Mt. Margaret, 27 Jan. 1963, *Verdcourt* 3573!;
Masai District: Laitokitok, 24 Feb. 1933, *C. G. Rogers* 488!
Tanzania. Arusha/Masai District: Engare Nanyuki R., 12 Dec. 1965, *Richards* 20819!;
Lushoto District: Amani, on hills beyond Monga Tea Estate, 17 Apr. 1968, *Renvoize
& Abdallah* 1541!; Uzaramo District: ± 9·5 km. E. of Ruvu R., 26 Nov. 1955,
Milne-Redhead & Taylor 7435!; Zanzibar I., Walezo, 18 Feb. 1930, *Vaughan* 1197!
Distr. **K**1–4, 6, 7; **T**2, 3, 5–8; **Z**; Ethiopia, Somalia, Angola, Zaire, Zambia, Malawi,
Mozambique, Rhodesia, South Africa (Natal), Madagascar
Hab. Woodland (usually deciduous) or grassland, forest clearings, old cultivations,
roadsides, usually in sandy or rocky soils; 0–2700 m.

Syn. *Lightfootia madagascariensis* A. DC., Monogr. Camp.: 116 (1830), *non Wahlenbergia madagascariensis* A. DC. Type: Madagascar, *Commerson* (P–JU, holo.!)
    L. abyssinica A. Rich., Tent. Fl. Abyss. 2: 1 (1851); Vatke in Linnaea 40: 200
      (1876); Hemsl. in F.T.A. 3: 474 (1877); Engl. in Abh. Preuss. Akad. Wiss.
      1891: 411 (1892); Brenan in Mem. N.Y. Bot. Gard. 8: 491 (1954); Adamson in
      Journ. S. Afr. Bot. 21: 209 (1955); E.P.A.: 1053 (1965), excl. syn. *L. abyssinica*
      var. *glaberrima* Engl.; U.K.W.F.: 511, 512 (fig.) (1974)
    L. abyssinica A. Rich. var. *tenuis* Oliv. in J.L.S. 21: 401 (1885). Type: Kenya,
      Naivasha District, Crater Lake, *J. Thomson* (K, holo.!)
    L. sodenii Engl. in E.J. 19, Beibl. 47: 52 (1894), as "*sodeni*". Types: Tanzania,
      Lushoto District, Mlalo, *Holst* 347 & 648 (B, syn. †)
    L. madagascariensis A. DC. var. *glabra* Engl., P.O.A. C: 400 (1895). Type:
      Tanzania, Uzaramo, *Stuhlmann* 6383 (B, holo. †, EA, lecto.!)
    L. rupestris Engl. in E.J. 30: 419 (1901). Type: Tanzania, Njombe District,
      Ukinga, Kipengere Range, *Goetze* 974 (B, holo. †, BM, lecto.!, BR, iso.!)
    L. divaricata Engl. in E.J. 32: 117 (1902); E.P.A.: 1054 (1965), *non* Buek (1837),
      nom. illegit. Type: Ethiopia, Harar, *Ellenbeck* 742 (B, holo. †)
    L. grandifolia Engl. in E.J. 40: 48 (1907). Types: Tanzania, E. Usambara Mts.,
      Sengerenu (?Sangerawe), *Engler* 881 & Tanzania, Pangani District, Makinyumbi
      [Masinjumbi], *Scheffler* 235 (B, syn. †)
    L. subulata Engl. in E.J. 40: 48 (1907), *non* L'Hérit. (1788), *nom. illegit.* Type:
      Kenya, Nakuru, *Engler* 2027 (B, holo. †)
    L. ellenbeckii Engl. in E.J. 40: 49 (1907); E.P.A.: 1054 (1965). Type: Ethiopia,
      Harar, *Ellenbeck* (B, holo. †)
    L. elata Chiov., Racc. Bot. Miss. Consol. Kenya: 74 (1935). Type: Kenya,
      Nyeri, *Balbo* 488 (TOM, lecto.!)
    L. arenicola Meikle in K.B. 1948: 466 (1949). Type: Zambia, Mwinilunga
      District, source of Matonchi Dambo, *Milne-Redhead* 2964 (K, holo.!, BM, BR,
      iso.)
    L. sp. A & L. sp. B sensu Agnew, U.K.W.F.: 511 (1974)

Note. The variation within the very widespread and polymorphic subsp. *abyssinica*
largely follows a more or less geographical pattern and locally it is admittedly often
possible to distinguish different forms. However, as the forms are so numerous
and vaguely defined and, in addition, intermediate or inconsistent specimens are

frequent I have refrained from a subdivision of this taxon. Instead I give a brief outline of the variation to be found within the Flora area and adjacent regions.

The coastal forms in Tanzania agree fairly well with coastal forms in Mozambique, Natal and Madagascar and have lax inflorescences, small leaves and sharply trigonous seeds (*Lightfootia madagascariensis, L. madagascariensis* var. *glabra*). The Madagascan material differs somewhat by having invariably hairy styles. This is an inconsistent character in corresponding forms in continental Africa and very rare in the Flora area.

Further inland in south-eastern Kenya, the Usambara Mts. and also in the Lake Malawi area there is a trend towards larger leaves and more contracted inflorescences (*L. grandifolia*). The same trend is to be found in Malawi, Mozambique and Rhodesia.

In southern Kenya and on Mt. Kilimanjaro, Pare Mts. and northern Usambara Mts. there is a stout form with densely hairy leaves and stems, ± lax inflorescences and glabrous styles (*L. sodenii*). Similar forms occur also in Ethiopia (including the type of *W. abyssinica*). Plants from central Kenya usually have contracted, ± spike-like inflorescences (*L. elata, L. sp. A* sensu Agnew), and may be superficially similar to *W. napiformis*, which is, however, easily distinguished by its compressed seeds and hairy styles.

Forms from the Mt. Meru area in Tanzania, mainly western Kenya and from parts of Ethiopia and Somalia are often glabrous or almost so, have small, narrow leaves, very lax inflorescences and ± obscurely trigonous seeds (*L. abyssinica* var. *tenuis, L. divaricata, L. subulata, L. ellenbeckii, L. sp. B* sensu Agnew).

Finally there is north of Lake Malawi a deviating form with decumbent stems and another with the flowers densely clustered terminally (*L. rupestris*). The latter is only known from the type collection in which there are no seeds.

subsp. **parvipetala** *Thulin* in Symb. Bot. Upsal. 21(1): 167, fig. 12/C, 19/A (1975). Type: Tanzania, Mbulu District, between Magugu and Babati, *Polhill & Paulo* 2372 (K, holo. !, BR, EA, LISC, iso. !)

Annual erect herb, up to 40 cm. tall, usually much branched from a whitish taproot. Leaves linear to lanceolate, up to 10–60 mm. long, 1·5–10 mm. wide. Calyx-lobes 0·8–2·2 mm. long. Corolla 2–3 mm. long, blue or lilac. Style glabrous below, very shortly 3-lobed (± 0·2 mm.) at the apex. Seeds trigonous, 0·4–0·5 mm. long.

Kenya. Machakos/Kitui District: Ukambani, 1892–93, *Gregory* 103 !; Kitui District: Nairobi–Garissa road, 5 km. E. of Ukazzi, 9 May 1974, *Gillett & Gachathi* 20500 !; Teita District: Tsavo West, Murka Camp, 16 Aug. 1967, *Gilbert* 1247 !
Tanzania. Kondoa District: Kandaga Scarp, 20 Jan. 1928, *B. D. Burtt* 1219 !; Bagamoyo District: ± 3 km. S. of Bagamoyo, 26 July 1970, *Thulin & Mhoro* 486 !; Rufiji District: Mafia, Chole I., 24 Sept. 1937, *Greenway* 5305 !
Distr. **K**4, 7; **T**1–3, 5, 6; not known elsewhere
Hab. In cultivated or waste ground, sandy places; 0–1650 m.

Syn. *Lightfootia tanneri* Agnew, U.K.W.F.: 511 (1974), *nom. nud.*

19. **W. collomioides** (*A. DC.*) *Thulin* in Symb. Bot. Upsal. 21(1): 168 (1975). Type: Angola, Huila, Morro de Lopollo, *Welwitsch* 1163 (G, holo. !, BM, BR, C, K, LISU, P, iso. !)

Annual or ? short-lived perennial herb, up to 1 m. tall. Stems erect, rarely unbranched, ± hirsute or glabrescent. Leaves sessile to subpetiolate, very narrowly elliptic to elliptic, 30–60 mm. long, 3–12(–17) mm. wide, acute, with cuneate base, ± hirsute; margin cartilaginous, undulate-dentate; upper leaves of ± the same size as the lower ones, forming a sort of involucre. Inflorescences terminal, dense, head-like, round or somewhat elongated; flowers subsessile or sometimes with glabrous or hairy pedicels up to 4 mm. long. Hypanthium obconical or hemispherical, glabrous or almost so, 5-nerved. Calyx-lobes 2–4 mm. long, glabrous or ciliate. Corolla 5–7 mm. long, white or blue, split almost to the base into lanceolate lobes, glabrous outside, ± puberulous inside near the base; tube 0·1–0·4 mm. long. Filament-bases dilated, very shortly and densely ciliate; anthers 1·5–3 mm. long. Ovary 3-locular, semi-inferior; style longer than the corolla, thickened in the upper part, eglandular, glabrous below; lobes 3, 0·6–0·8 mm. long. Capsule 3-locular, 5-nerved; valves ± 1 mm. long. Seeds elliptic in outline with acute ends, trigonous, 0·5–0·7 mm. long; testa almost smooth.

TANZANIA. Songea District: 3 km. W. of Songea, 30 Apr. 1956, *Milne-Redhead & Taylor* 9958!; Tunduru District: 19 km. W. of Tunduru, 12 May 1962, *Boaler* 594!; Lindi District: Nachingwea, 22 May 1955, *Anderson* 1055!
DISTR. **T**4, 6, 8; Central African Republic, Congo (Brazzaville), Zaire, Angola, Malawi, Zambia, Mozambique
HAB. Deciduous woodland, weed on cultivated ground, usually in sandy or stony soils; 200–1700 m.

SYN. *Lightfootia collomioides* A. DC. in Ann. Sci. Nat., sér. 5, 6: 328 (1866); Hemsl. in F.T.A. 3: 475 (1877); Hiern, Cat. Afr. Pl. Welw. 1: 630 (1898); Engl. & Gilg in Kun.-Samb.-Exped.: 396 (1903); Lambinon & Duvign. in B.S.B.B. 93: 48 (1961)
    L. leptophylla C. H. Wright in Hook., Ic. Pl. 27, t. 2659 (1900); Lambinon & Duvign. in B.S.B.B. 93: 52 (1961). Type: Mozambique, Niassa, between Unangu and Lake Shire, *Johnson* 40 (K, holo.!)
    [*L. napiformis* sensu De Wild., Pl. Bequaert. 2: 124 (1923), *non* A. DC.]
    L. collomioides A. DC. subsp. *katangensis* Lambinon in B.S.B.B. 93: 49, fig. 3 (1961). Type: Zaire, Katanga, Mitonte, *Duvigneaud* 3005 L (BRLU, holo.!)

NOTE. All plants from the Flora area and from Mozambique are annuals with often distinct pedicels and have usually been called *Lightfootia leptophylla*. However, the variation is clearly clinal and ± perennial forms with subsessile flowers become increasingly common westwards in the range. All annual specimens seen from Zambia and Zaire have subsessile flowers.

20. **W. polycephala** (*Mildbr.*) *Thulin* in Symb. Bot. Upsal. 21(1): 173, fig. 9/L, 12/L, 28/B (1975). Type: Tanzania, Iringa District, Lupembe, Ruhudji R., *Schlieben* 805 (B, holo.!, BM, BR, G, M, P, S, Z, iso.!)

Perennial prostrate or decumbent herb from a woody taproot. Stems many, trailing, up to 1·2 m. long, hirsute, branching at a ± wide angle. Leaves sessile, lanceolate, 5–17 mm. long, 1–5 mm. wide, acute, with cuneate to truncate base, hirsute; margin cartilaginous, undulate-dentate, sometimes only sparsely so. Inflorescences usually many terminal dense spherical heads, ± 1 cm. in diameter, occasionally also with a few lateral flowers; flowers subsessile. Hypanthium obconical, glabrous or almost so, obscurely 10-nerved. Calyx-lobes 1·5–3 mm. long, ± hirsute. Corolla 3·5–4(–5·5) mm. long, white, split almost to the base into lanceolate lobes, glabrous outside, ± hairy inside near the base. Filament-bases angularly obovate, ciliate; anthers 1·5–2·4 mm. long. Ovary 3-locular, semi-inferior; style about as long as the corolla, not thickened in upper part, eglandular, glabrous below; lobes 3, ± 0·4 mm. long. Capsule 3-locular, obscurely 10-nerved; valves ± 0·8 mm. long. Seeds elliptic in outline with acute ends, trigonous, 0·7–0·8 mm. long; testa almost smooth, brown (often darkly so).

TANZANIA. Iringa District: near Ifunda, 22 Sept. 1970, *Thulin & Mhoro* 1088!; Njombe District: Njombe, Oct. 1931, *Staples* 166! & ± 12 km. N. of Songea/Njombe District Boundary, 6 July 1956, *Milne-Redhead & Taylor* 10762!
DISTR. **T**7; not known elsewhere
HAB. Upland grassland or woodland; 1550–1800 m.

SYN. *Lightfootia polycephala* Mildbr. in N.B.G.B. 11: 686 (1932)
    Lightfootia glomerata Engl. var. *capitata* (Bak.) Lambinon in B.S.B.B. 93: 47 (1961), pro parte

21. **W. hirsuta** (*Edgew.*) *Tuyn* in Fl. Males., ser. 1, 6: 113 (1960); Thulin in Symb. Bot. Upsal. 21(1): 174, fig. 11/N, 12/H, 27/D, 29/A-E (1975). Type: India, Himalaya, Banásar, *Edgeworth* 252 (K, holo.!)

Annual ± erect herb, 4–30 cm. tall. Stem usually with numerous widely spreading branches, ± hirsute. Leaves sessile to subpetiolate, narrowly oblanceolate to obovate or broadly elliptic to ovate, 10–55 mm. long, 4–25 mm. wide, obtuse to subacute, rarely almost acuminate, with attenuate to cuneate base, ± hirsute; margin cartilaginous, undulate-dentate; midvein

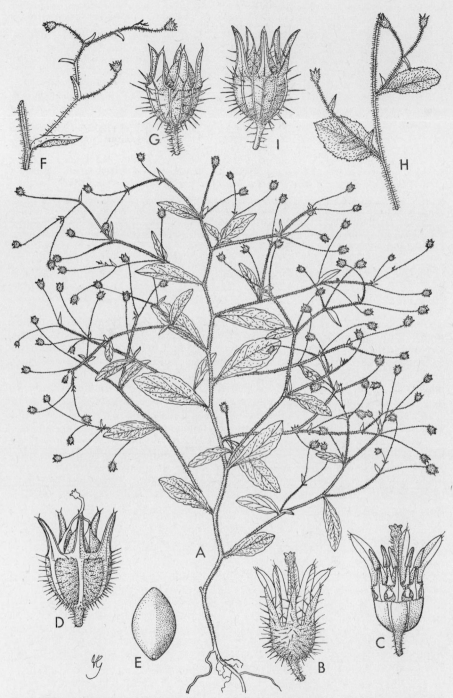

FIG. 5. *WAHLENBERGIA HIRSUTA*—**A,** habit, × ⅔; **B,** flower, × 8; **C,** same, with calyx-lobes, two petals and pubescence of hypanthium removed, × 8; **D,** capsule, × 6; **E,** seed, × 30. *W. ERECTA*—**F,** flowering branch, × 1; **G,** capsule, × 6. *W. FLEXUOSA*—**H,** flowering branch, × 1; **I,** capsule, × 6. A, from *Renvoise* 2238; B–E, from *Polhill & Paulo* 2019; F, G, from *B. D. Burtt* 3606; H, I, from *Letouzey* 7339.

and lateral veins prominent beneath. Inflorescence lax, spreading; pedicels 5–20 mm. long, with very short but usually dense pubescence, often mixed with longer hairs. Hypanthium hemispherical, 5-nerved, ± hirsute. Calyx-lobes 1–3(–5) mm. long. Corolla 1·5–4 mm. long, blue to pale blue or white, split almost to the base into lanceolate lobes, glabrous or hairy outside. Stamens with filament-bases ± rhombic, ciliate; anthers 0·5–1 mm. long. Ovary 3-locular, semi-inferior; style about as long as or longer than the corolla, thickened in the upper part, eglandular, subcapitate or shortly 3-lobed at the apex, glabrous or hairy below. Capsule 3-locular, 5-nerved; valves 3, 1–2·5 mm. long. Seeds elliptic in outline with acute ends, trigonous, 0·4–0·6 mm. long; testa almost smooth, brown or sometimes almost black. Fig. 5/A–E.

UGANDA. Karamoja District: Napak Mts., June 1964, *J. Wilson* 1632; Bunyoro District: Masindi, 4 Jan. 1962, *Turner* 511 T; Mengo District: Kivuvu, Nov. 1914, *Dummer* 1224!

KENYA. Trans-Nzoia District: Kitale, 10 Sept. 1956, *Bogdan* 4336! & Kitale, Nov. 1967, *Tweedie* 3508!; Uasin Gishu District: near Turbo, *Brodhurst Hill* 457!

TANZANIA. Mbulu District: Pienaars Heights, 4 May 1962, *Polhill & Paulo* 2326!; Njombe District: Iyayi, 15 Apr. 1962, *Polhill & Paulo* 2019!; Songea District: Kitai, 17 Apr. 1956, *Milne-Redhead & Taylor* 9747!

DISTR. U1, 2, 4; K3, 5; T2, 4–8; Senegal, Ghana, Nigeria, Cameroun, Central African Republic, Sudan, Ethiopia, Angola, Zaire, Rwanda, Burundi, Zambia, Malawi, Rhodesia, Comoro Is., Madagascar, India, Nepal

HAB. Grassland or woodland, roadsides, waste or cultivated ground; 900–1850 m.

SYN. *Cephalostigma hirsutum* Edgew. in T.L.S. 20: 81 (1846)
    [*C. erectum* sensu Vatke in Linnaea 38: 699 (1874), quoad syn. pro parte; Engl. in Abh. Preuss. Akad. Wiss. 1891: 408 (1892), pro parte; Brenan in Mem. N.Y. Bot. Gard. 8: 490 (1954), pro parte]
    C. erectum (Roem. & Schultes) Vatke var. *coeruleum* Chiov. in Ann. Bot. Roma 9: 79 (1911). Type: Ethiopia, Beghemder, Dembia, near Asoso, *Chiovenda* 2015 (FI, lecto. !)
    [*Cephalostigma perotifolium* sensu Hutch. & Dalz., F.W.T.A. 2: 191 (1931), pro parte]
    Lightfootia hirsuta (Edgew.) Hepper in K.B. 15: 61 (1961), pro parte & in F.W.T.A., ed. 2, 2: 311 (1963), pro parte; U.K.W.F.: 511 (1974)
    [*Wahlenbergia erecta* sensu E.P.A.: 1054 (1965), quoad syn. pro parte, *non* (Roem. & Schultes) Tuyn]

NOTE. *W. hirsuta* has often been confused with *W. erecta* and *W. flexuosa*, but is easily distinguished by its trigonous seeds and 5-nerved hypanthium. Also the pubescence of *W. hirsuta* is characteristic, especially on the pedicels, with numerous short often papillary hairs usually mixed with longer hairs.

22. **W. hookeri** (*C. B. Cl.*) *Tuyn* in Fl. Males., ser. 1, 6: 114, fig. 1/a–g (1960); Thulin in Symb. Bot. Upsal. 21(1): 181, fig. 14/K (1975). Type: India, Bihar, Chota Nagpore, *Clarke* 24796 (K, lecto. !, BM, iso. !)

Annual erect herb, 4–25 cm. tall. Stem much branched at least above the leaves, which tend to be verticillate at the middle of the stem, hirsute at the base. Leaves sessile to subpetiolate, lanceolate to narrowly ovate, 8–35(–48) mm. long, 3–10(–13) mm. wide, acute to subacute with cuneate base, ± hirsute; margin cartilaginous, shallowly undulate-dentate; midvein prominent beneath as are the lateral veins. Inflorescence lax, spreading; pedicels up to 7–20(–25) mm. long, glabrous. Hypanthium hemispherical, 5-nerved, glabrous. Calyx-lobes 0·5–1·2 mm. long. Corolla 1·2–2 mm. long, blue to pale violet, split almost to the base into linear lobes. Filament-bases narrowly dilated, ciliate; anthers ± 0·4 mm. long. Ovary 2-locular, subinferior; style about as long as the corolla, eglandular, glabrous below, subcapitate at the apex with 2 very short lobes. Capsule 2-locular, 5-nerved; valves ± 1 mm. long. Seeds elliptic in outline with acute ends, trigonous, 0·4–0·6 mm. long; testa almost smooth. Fig. 2/K, p. 6.

Kenya. Trans-Nzoia District: Kitale, 2 Oct. 1959, *Verdcourt* 2460!; Nandi District: Kiptuiya, 19 Sept. 1960, *Tallantire* 124!
Tanzania. Lushoto District: W. Usambara Mts., Sakare, 24 Sept. 1902, *Engler* 979 in part!
Distr. K3; T3; Cameroun, Ethiopia, Zaire, India, Thailand, Java
Hab. Grassland or woodland, often in rocky places or on cultivated ground; 800–1950 m.

Syn. *Cephalostigma hookeri* C. B. Cl. in Fl. Brit. Ind. 3: 429 (1881)

23. **W. ramosissima** (*Hemsl.*) *Thulin* in Symb. Bot. Upsal. 21(1): 187 (1975). Type: Cameroun, Mt. Cameroun, *Mann* 1333 (K, lecto.!)

Annual ± erect herb, 10–50 cm. tall. Stem ± hirsute or rarely glabrous. Leaves sessile, basal ones ± ovate, caducous, stem leaves linear to elliptic, 5–30(–50) mm. long, 1–5(–9) mm. wide, acute to subacute with attenuate to truncate base, ± hirsute or glabrous; margin cartilaginous, usually flat or slightly undulate, ± denticulate; midvein prominent beneath, lateral veins ± obscure. Inflorescence lax; pedicels (1–)3–30 mm. long, glabrous or rarely hirsute. Hypanthium obconical or obovoid to hemispherical, 5–10-nerved, glabrous or ± hirsute. Calyx-lobes 0·5–2·5 mm. long, entire or sparsely denticulate, glabrous or ± hairy. Corolla 1·2–5 mm. long, white to blue or yellowish, split almost to the base into linear lobes, glabrous or hirsute outside.   Filament-bases ± dilated, ciliate; anthers rounded to elongated, 0·3–1·6 mm. long. Ovary 2-locular, semi-inferior; style about as long as or somewhat shorter than the corolla, eglandular, glabrous or hairy below, slightly thickened to subcapitate in the upper part; lobes 2, very short to distinct. Capsule 2-locular, 5–10-nerved; valves 2, 0·5–1·2 mm. long. Seeds elliptic in outline with acute ends, trigonous, 0·4–0·6 mm. long; testa almost smooth.

Syn. *Cephalostigma ramosissimum* Hemsl. in F.T.A. 3: 472 (1877); Hutch. & Dalz., F.W.T.A. 2: 191 (1931)
         Lightfootia ramosissima (Hemsl.) Hepper in K.B. 15: 61 (1961) & in F.W.T.A., ed. 2, 2: 311 (1963)

Key to infraspecific variants

Style subcapitate at the apex, densely hairy below:
  Leaves narrowly elliptic, up to 5 mm. wide   .   .   subsp. **subcapitata**
  Leaves linear, up to 2(–3) mm. wide   .   .   .   subsp. **centiflora**
Style distinctly 2-lobed at the apex, glabrous below   .   subsp.
                                             oldenlandioides

subsp. **subcapitata** *Thulin* in Symb. Bot. Upsal. 21(1): 192, fig. 32/D (1975). Type: Tanzania, Songea District, Matengo Hills, 11 km. N. of Miyau, *Milne-Redhead & Taylor* 10410 (K, holo.!, BR, EA, LISC, iso.!)

Stem branching also from near the base, up to 40 cm. tall, hirsute at least below. Leaves narrowly elliptic, up to 17 mm. long and 4–5 mm. wide, acute or subacute, ± hirsute. Pedicels up to 30 mm. long, glabrous or almost so. Hypanthium 5-nerved, somewhat hirsute. Calyx-lobes up to 1·2 mm. long. Corolla blue with white centre, up to 2·8 mm. long, glabrous. Anthers ± 0·6 mm. long. Style subcapitate, very shortly lobed at the apex, densely hairy below.

Tanzania. Songea District: Matengo Hills, 11 km. N. of Miyau, 22 May 1956, *Milne-Redhead & Taylor* 10410!
Distr. T8; Malawi
Hab. Derelict cultivated land

Note. Three collections from Katanga cited by me (loc. cit.) under this subspecies are probably better regarded as small-leaved forms of *W. perrottetii*. The distinction between *W. ramosissima* and *W. perrottetii* is not sharp and subsp. *subcapitata* is near the borderline, mainly distinguished by its smaller acute leaves with only slightly undulate margins.

subsp. **centiflora** *Thulin* in Symb. Bot. Upsal. 21(1): 194, fig. 32/F, 34/A (1975). Type: Zambia, D'hulmiti–Katula on old Mbala–Mpulungu road, *Richards* 5565 (K, holo.!, BR, iso.!)

Stem usually much branched in upper part only, 10–30 cm. tall, ± hirsute below. Leaves linear to narrowly lanceolate, up to 12–30 mm. long, 1–2(–3) mm. wide (basal leaves occasionally up to 5 mm.). Pedicels 3–23 mm. long, glabrous. Hypanthium 5-nerved, glabrous or occasionally hairy. Calyx-lobes 0·5–1 mm. long. Corolla 2–3·2 mm. long, blue to white. Anthers 0·4–0·6(–0·8) mm. long. Style subcapitate, very shortly lobed at the apex, usually densely hairy below. Seeds ± 0·4 mm. long.

Tanzania. Ufipa District: Kalambo Falls, 10 Apr. 1950, *Bullock* 2852!; Iringa District: Malangali, 28 Mar. 1962, *Polhill & Paulo* 1907!
Distr. **T**4, 7; Zaire, Zambia, Malawi, Angola
Hab. Deciduous woodland or grassland, usually in moist places in sandy or rocky soils; 900–1650 m.

subsp. **oldenlandioides** *Thulin* in Symb. Bot. Upsal. 21(1): 196, fig. 32/G, 33/A (1975). Type: Tanzania, Iringa District, Malangali, *Polhill & Paulo* 1906 (K, holo.!, BR, EA, LISC, P, iso.!)

Stem much branched in upper part only, 4–10 cm. tall, hirsute at the base. Leaves linear (basal leaves ovate), 5–15 mm. long, up to 1·5 mm. wide, acute. Pedicels up to 15 mm. long, glabrous. Hypanthium 5-nerved, hirsute. Calyx-lobes 0·7–0·9 mm. long. Corolla ± 1·2 mm. long, white. Anthers ± 0·3 mm. long, rounded. Style shortly but distinctly 2-lobed at the apex, glabrous below. Seeds ± 0·4 mm. long.

Tanzania. Iringa District: Malangali, 29 Mar. 1962, *Polhill & Paulo* 1906! & ± 15 km. E. of Madibira, 29 Sept. 1970, *Thulin & Mhoro* 1270!
Distr. **T**7; Zambia
Hab. *Brachystegia* woodland on red sandy or stony soils; 1650–1800 m.

Note. Six further subspecies are recognized by me (loc. cit.), of which subsp. *ramosissima* occurs in Nigeria and Cameroun, subsp. *A* in Angola, subsp. *zambiensis* in Zambia and Zaire, subsp. *lateralis* in Angola, Rhodesia, Botswana and South West Africa, subsp. *richardsiae* in Zambia and subsp. *B* in Zaire. Subsp. *zambiensis*, with 10-nerved hypanthium, hirsute corolla, narrowly elliptic or lanceolate leaves and stem branching also from near the base, and subsp. *richardsiae*, which is glabrous with linear leaves and a distinctly 2-lobed style, glabrous below, may well be present also in the Flora area.

24. **W. perrottetii** (*A. DC.*) *Thulin* in Symb. Bot. Upsal. 21(1): 199, fig. 10/M, 12/K, 14/F, O, 17/C (1975). Type: Senegal, Kham, *Leprieur & Perrottet* (G-DC, holo.!)

Annual erect herb, 0·2–0·6(–1) m. tall. Stem hirsute at the base. Leaves narrowly elliptic to lanceolate, up to 18–57 mm. long, 5–16 mm. wide, ± obtuse and mucronulate, with cuneate base, hirsute at least on the lower leaves; margin cartilaginous, undulate-crenate; midvein and lateral veins prominent beneath. Inflorescence lax; pedicels up to 25 mm. long, glabrous or hirsute. Hypanthium obconical, 5-nerved, glabrous or hirsute. Calyx-lobes 0·7–1·6 mm. long. Corolla 1·8–3 mm. long, white to blue, split almost to the base into linear-lanceolate lobes, glabrous or with a few hairs outside. Filament-bases dilated, ciliate; anthers 0·6–0·9 mm. long. Ovary 2-locular, semi-inferior; style as long as or longer than the corolla, often blue, eglandular, hairy below, subcapitate with 2 very short lobes at the apex. Capsule 2-locular, 5-nerved, glabrous or hirsute; valves 2, ± 1 mm. long. Seeds elliptic in outline with acute ends, trigonous, 0·5–0·6 mm. long; testa almost smooth. Fig. 2/F, O, p. 6.

Uganda. W. Nile District: Arua, Dec. 1937, *Hazel* 414!; Kigezi District: Bufumbira, June 1947, *Purseglove* 2457!
Tanzania. Buha District: Kakombe, 6 July 1959, *Newbould & Harley* 4254!; Iringa District: 12 km. W. of Kidatu, 4 Sept. 1970, *Thulin & Mhoro* 858!
Distr. **U**1, 2; **T**4, 7; western tropical Africa from Senegal to Angola, Zaire, Burundi, Malawi, Madagascar, Comoro Is., South America

HAB. Grassland, woodland, waste places, or as a weed on cultivated ground; 450–1950 m.

SYN. *Cephalostigma perrottetii* A. DC., Monogr. Camp.: 118 (1830); Hemsl. in F.T.A. 3: 472 (1877); De Wild., Pl. Bequaert. 4: 425 (1928); Hutch. & Dalz., F.W.T.A. 2: 191 (1931); Hepper in F.W.T.A., ed. 2, 2: 311 (1963), pro parte
  C. prieurii A. DC., Monogr. Camp.: 118 (1830); Hemsl. in F.T.A. 3: 473 (1877), as " *C. prieurei* ". Type: Senegal, Jonal, *Perrottet* (G-DC, holo. !)

NOTE. *W. perrottetii* is close to *W. ramosissima*, and I treat it as a distinct species mainly because of its relative uniformity over a very large area of distribution, in contrast to the forms included in *W. ramosissima* which are all more or less local. The same consideration is made for the next species, *W. paludicola*. Although very unlike *W. perrottetii*, *W. paludicola* is also close to some of the forms in the polymorphic *W. ramosissima*.

25. **W. paludicola** *Thulin* in Symb. Bot. Upsal. 21(1): 202, fig. 35 A (1975). Type: Zambia, Bangweulu, Kamindas, *R. E. Fries* 891 (UPS, lecto. !)

Annual slender ± trailing or straggling herb, 10–50 cm. tall, glabrous. Stem usually branching only in upper part, furrowed. Leaves sessile; basal ones ± ovate, caducous; stem leaves linear to narrowly lanceolate, 2–12 mm. long, 0·2–1(–1·6) mm. wide, acute; margin cartilaginous with a few denticles; midvein prominent beneath, lateral veins obscure. Inflorescence lax; pedicels 6–32 mm. long. Hypanthium 5(–10)-nerved, glabrous. Calyx-lobes 0·6–1·4 mm. long. Corolla 2·4–3·5 mm. long, white or bluish, split almost to the base into lanceolate lobes. Filament-bases narrowly dilated, ± ciliate; anthers 0·35–0·8 mm. long. Ovary 2-locular, semi-inferior; style about as long as or somewhat shorter than the corolla, eglandular, glabrous or ± hairy below, slightly thickened in the upper part, subcapitate and with 2 very short lobes at the apex. Capsule 2-locular, 5(–10)-nerved; valves 2, 0·6–1·4 mm. long. Seeds elliptic in outline with acute ends, trigonous, 0·45–0·6 mm. long; testa almost smooth.

UGANDA. Masaka District: Lake Nabugabo, June 1937, *Chandler* 1706!; Mengo District: Salama, Dec. 1917, *Dummer* 3903!
TANZANIA. Iringa District: Mufindi, 21 Dec. 1969, *Paget-Wilkes* 718!; Songea District: 6·5 km. E. of Gumbiro, 8 May 1956, *Milne-Redhead & Taylor* 10015!; Tunduru District: 1·5 km. E. of Muhuwezi, 21 Dec. 1955, *Milne-Redhead & Taylor* 7737!
DISTR. U4; T7, 8; Cameroun, Zaire, Zambia, Malawi, Rhodesia, Angola, Madagascar
HAB. Boggy grassland, often on sandy soils; 450–950 m. (in Malawi up to 2200 m.)

SYN. *Lightfootia gracillima* R. E. Fries, Schwed. Rhod.-Kongo-Exped. 2: 316 (1916), non *Wahlenbergia gracillima* S. Moore. Type as for *Wahlenbergia paludicola*
  [*L. abyssinica* A. Rich. var. *tenuis* sensu Hepper in K.B. 15: 61 (1961) & in F.W.T.A., ed. 2, 2: 311 (1963), *non* Oliv.]

3. GUNILLAEA

Thulin in Bot. Notiser 127: 166 (1974)

Annual herbs. Leaves alternate, sessile, flat. Inflorescences ± leafy, monochasial. Flowers ± regular, protandrous. Calyx-lobes (3–)4–5, accrescent. Corolla campanulate, (3–)4–5-lobed. Stamens (3–)4–5, free; filament-bases almost linear to broadly dilated, ciliate or glabrous. Ovary inferior, 2-locular; ovules numerous; style shorter than the corolla, 2-lobed, upper part with pollen-collecting hairs, lower part glabrous or hairy; 2 glands sometimes present at the base of the style-lobes. Capsule thin-walled, indehiscent, tardily opening by the irregular decomposition of the pericarp between the persistent lateral nerves. Seeds numerous; testa sulcate, sometimes with hair-like projections.

Two species in tropical Africa, one extending to Madagascar.

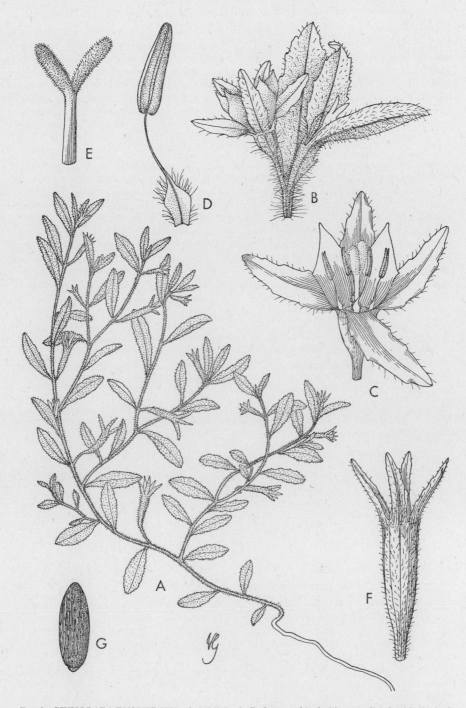

FIG. 6. *GUNILLAEA EMIRNENSIS*—**A,** habit, × ⅔; **B,** flower and leaf with sympodial shoot in the leaf-axil, × 4; **C,** flower with two petals removed, × 6; **D,** stamen, × 18; **E,** style, × 12; **F,** capsule, × 3; **G,** seed, × 54. All from *Robson* 667.

G. emirnensis (*A. DC.*) *Thulin* in Bot. Notiser 127 : 166, fig. 1, 3/F–K, 10/C, F (1974). Type : Madagascar, Emirna, *Bojer* (G-DC, holo. !)

Annual ascending or decumbent, rarely erect herb, 4–40 cm. tall, usually much branched from the base. Stem ± hirsute, rarely glabrescent. Leaves narrowly elliptic to elliptic or oblanceolate, up to 8–40 mm. long and 2–10 mm. wide, acute to almost rounded at the apex, with attenuate base, ± hirsute or glabrous; margin slightly cartilaginous, ± undulate-crenate; midvein protruding beneath, lateral veins obscure. Inflorescence not well demarcated, lax. Flowers sessile or shortly pedicellate; pedicel elongating in fruit up to 10 mm. Hypanthium narrowly obconical, 5–10-nerved, glabrous or hirsute. Calyx-lobes 3–5, often of varying length, narrowly triangular to oblanceolate or narrowly oblong, 1·5–5(–10) mm. long, glabrous or ± hirsute, sparsely denticulate. Corolla white or blue, 2·4–4(–5) mm. long; lobes 3–4(–5), united about halfway. Stamens 3–4(–5), 2–3·2 mm. long; filament-bases narrowly triangular to broadly angular-obovate, glabrous or ciliate; anthers 0·6–1·6 mm. long. Style glabrous or rarely hairy below, eglandular; lobes 2, 0·5–1·6 mm. long, often slightly unequal in length; pollen-collecting hairs present on style-lobes only. Capsules obovoid to narrowly obconical or cylindrical, often slightly curved upwards, up to 10 mm. long, prominently 5–10-nerved, glabrous or hirsute. Seeds elliptic, often ± reniform, slightly compressed, 0·5–0·8 mm. long; testa sulcate, rarely with hair-like projections. Fig. 6.

TANZANIA. Ulanga District : Taveta, Dec. 1959, *Haerdi* 403/0 !
DISTR. **T6**; Angola, Zaire, Zambia, Malawi, Rhodesia, Madagascar
HAB. Sand or mud in ± wet places such as river banks, lake shores, swamps or rice fields; 550 m. (up to 2150 m. outside East Africa)

SYN. *Wahlenbergia emirnensis* A. DC. in Prodr. 7 : 432 (1839)
       W. huillana A. DC. in Ann. Sci. Nat., sér., 5, 6 : 333 (1866); Hemsl. in F.T.A. 3 : 479 (1877). Type : Angola, Huila, Lopollo–Lake Ivantâla, *Welwitsch* 1161 (G, holo. !, BM, BR, C, K, LISU, M, P, iso. !)
       W. huillana A. DC. var. *pusilla* A. DC. in Ann. Sci. Nat., sér. 5, 6 : 333 (1866). Type : Angola, Huila, Empalanca, *Welwitsch* 1160 (G, holo. !, BM, LISU, iso. !)
       Cervicina huillana (A. DC.) Hiern, Cat. Afr. Pl. Welw. 1 : 631 (1898)

4. CAMPANULA

L., Sp. Pl. : 163 (1753) & Gen. Pl., ed. 5 : 77 (1754); A. DC., Monogr. Camp. : 213 (1830)

Annual, biennial or perennial herbs, rarely subshrubs, glabrous or hairy. Leaves alternate, simple. Inflorescences panicle-, raceme-, spike- or head-like, or flowers solitary. Calyx-lobes usually 5, sometimes with reflexed appendages in between them. Corolla campanulate to almost rotate or cylindrical, ± deeply 5-lobed. Stamens 5, free; filaments usually dilated and ciliate at the base. Ovary ± inferior, 3–5-locular; ovules numerous; style eglandular, with pollen-collecting hairs in the upper part, lower part glabrous or hairy; lobes 3–5. Capsule dehiscing by lateral pores or valves. Seeds numerous, ± elliptic in outline.

Some 300 species, widely distributed in the northern hemisphere, especially abundant in the Mediterranean region and the Middle East. The East African species are the only ones extending south of the equator. Many species are of horticultural value.
Campanula rapunculoides L., a species widespread in Europe and Asia, has been found introduced in Kenya (Muguga, Kagia Farm, *Dyer* !). It might well become established as a weed in the highlands.

Plant perennial; corolla with rather long stiff hairs on
   midveins of lobes or glabrous; anthers 1·5–3 mm.
   long   .   .   .   . 1. *C. edulis*

Plant annual; corolla densely puberulous all over outer
surface, usually also with longer hairs on midveins of
lobes; anthers 0·8–1·5 mm. long . . . . 2. *C. keniensis*

1. **C. edulis** *Forssk.*, Fl. Aegypt.-Arab.: 44 (1775): A. DC., Monogr. Camp.:
235 (1830); F.P.S. 3: 70 (1956); E.P.A.: 1052 (1965); Thulin in Bot. Notiser
128: 354, fig. 1/B (1976). Type: Yemen, Kurma, *Forsskål* (C, lecto. !)

Perennial; stems 6–70 cm. long, few to many, decumbent or ± erect,
ribbed, hirsute. Leaves sessile, elliptic to oblanceolate, up to 10–50 mm.
long, 3–16 mm. wide, acute to rounded at the apex, with attenuate to sub-
auriculate base; margin entire to ± distantly and obscurely dentate, sometimes somewhat undulate, hirsute especially on the veins. Inflorescence
lax, ± leafy, sometimes reduced to a single terminal flower. Flowers ± erect
on hirsute, up to 4 cm. long pedicels. Hypanthium broadly obconical, ±
10-nerved, hirsute. Calyx-lobes lanceolate to narrowly triangular, 3·5–
12 mm. long, acute, ciliate-pubescent with triangular reflexed appendages.
Corolla blue or purple with whitish base, rarely all white, campanulate to
almost cylindrical, 7–25(–30) mm. long, with corolla-lobes 1·5–11 mm. long,
often pilose on the veins outside, otherwise glabrous. Stamens with broadly
triangular ciliate filament-bases; anthers 1·5–3 mm. long. Ovary 3(–5)-
locular; style shorter than the corolla, narrowly 3–5-lobed, glabrous or hairy
at the base. Capsule nodding, 3(–5)-locular, dehiscing by small basal valves.
Seeds numerous, elliptic-oblong in outline, compressed, 0·7–0·8 mm. long,
almost smooth, yellowish brown. Fig. 7, p. 36.

UGANDA. Karamoja District: Mt. Moroto, Imagit Peak, Sept. 1956, *J. Wilson* 263!
& 6 Sept. 1956, *Hardy & Bally* in *Bally* 10770!; Kigezi District: Mt. Muhavura, N.
slopes, 11 Jan. 1933, *C. G. Rogers & Gardner* 341!
KENYA. Northern Frontier Province: Marsabit, crater above Boma, 18 Oct. 1947,
Bally 5670!; Aberdare Mts., Kinangop, Lereko, July 1932, *Napier* 2107! & S.
Kinangop, 21 May 1968, *J. Williams* WB2!
TANZANIA. Masai District: S. rim of Ngorongoro Crater, 21 June 1965, *Herlocker* 136!;
Mbulu District: Mt. Hanang, Werther's Peak, 12 Feb. 1946, *Greenway* 7709!; Arusha
District: Mt. Meru, E. side, 7 Apr. 1965, *Richards* 20086!
DISTR. U1, 2; K1, 3, 4, 6; T2; Ethiopia, Sudan, Somalia, Rwanda and Yemen, possibly
also in Chad and S. Algeria
HAB. Upland grassland, often on rocky ground; 1600–3700 m.

SYN. *C. esculenta* A. Rich., Tent. Fl. Abyss. 2: 4 (1851). Type: Ethiopia, Tigre,
Ouodgerate, *Quartin Dillon & Petit* (P, lecto. !)
C. quartiniana A. Rich., Tent. Fl. Abyss. 2: 5 (1851); Hemsl. in F.T.A. 3: 481
(1877); E.P.A. 1053 (1965). Type: Ethiopia, Tigre, Memsah, *Quartin Dillon
& Petit* (P, holo. !)
C. rigidipila A. Rich., Tent. Fl. Abyss. 2: 3 (1851); Hemsl. in F.T.A. 3: 482;
U.K.W.F.: 509 (1974). Type: Ethiopia, Tigre, Ouodgerate, *Quartin Dillon &
Petit* (P, lecto. !)
C. sarmentosa A. Rich., Tent. Fl. Abyss. 2: 4 (1851). Type: Ethiopia, Choa,
Quartin Dillon & Petit (P, lecto. !)
C. schimperi Vatke in Linnaea 38: 712 (1874), *nom. superfl.* Type as for *C.
rigidipila*
C. schimperi Vatke var. *quartiniana* (A. Rich.) Vatke in Linnaea 40: 201 (1876),
nom. illegit.
C. schimperi Vatke var. *rigidipila* (A. Rich.) Vatke in Linnaea 40: 201 (1876),
nom. illegit.
C. schimperi Vatke var. *sarmentosa* (A. Rich.) Vatke in Linnaea 40: 201 (1876),
nom. illegit.
C. rigidipila A. Rich. var. *quartiniana* (A. Rich.) Engl. in Abh. Preuss. Akad.
Wiss. 1891: 410 (1892)
C. rigidipila A. Rich. var. *sarmentosa* (A. Rich.) Engl. in Abh. Preuss. Akad. Wiss.
1891: 410 (1892)
C. rigidipila A. Rich. var. *esculenta* (A. Rich.) Di Capua in Ann. Ist. Bot. Roma 8:
236 (1904)

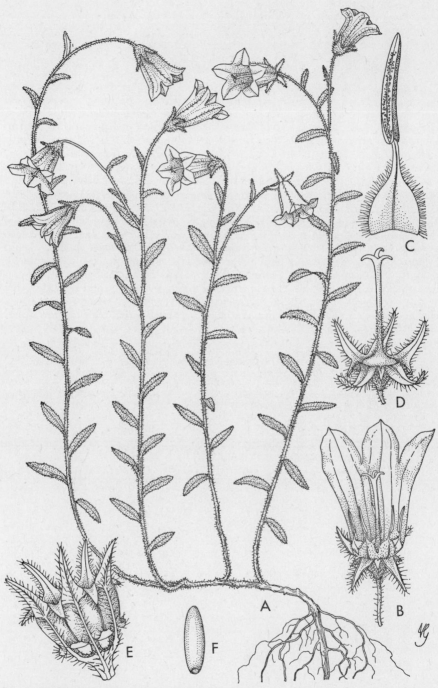

Fig. 7. *CAMPANULA EDULIS*, large-flowered form—**A**, habit, × ⅔; **B**, flower with two petals removed, × 1½; **C**, stamen, × 6; **D**, flower with corolla and stamens removed, × 2; **E**, capsule after dehiscence, × 3; **F**, seed, × 24. A, from *Eggeling* 2881 & *Herlocker* 136; B–D, from *Eggeling* 2881; E–F, from *Mabberley* 610.

Note. The species is very variable especially in flower-size. The variation is continuous, however, even though large-flowered forms predominate on certain mountains, e.g. Mt. Moroto in Uganda, Mts. Hanang and Ngorongoro in Tanzania and some others in Ethiopia.

2. **C. keniensis** *Thulin* in Bot. Notiser 128 : 350, fig. 1/A, 2/D (1976). Type : Kenya, Masai District, Ngong Hills, *Moberg* 1415 (UPS, holo. !)

Annual ± stiffly erect herb, up to 35 cm. tall. Stem branched mainly in the upper part, strongly ribbed, hirsute with mixed hairs of very variable lengths. Leaves sessile, narrowly ovate to ovate above, elliptic to oblanceolate or narrowly ovate towards the base, up to 10–25 mm. long, 5–10 mm. wide, acute or subacute, with truncate or at least in upper leaves, cordate base, hirsute ; margin ± undulate-crenate ; midvein and lateral veins prominent beneath. Inflorescence lax, leafy, with marked overtopping of the terminal flower giving a dichotomous appearance ; pedicels short, elongating up to 10 mm. in fruit. Hypanthium broadly obconical, with 5 distinct nerves and up to 5 additional ± weak nerves in between them, shortly and densely pubescent but with long hairs on the nerves. Calyx-lobes narrowly triangular, 4–8 mm. long, acute, with long hairs at margins and on midvein outside, otherwise shortly and densely pubescent on both sides ; calyx-appendages ovate, 1·5–2·5 mm. long, reflexed, ± obtuse. Corolla blue or mauve with whitish base, cylindrical, 6–8 mm. long, with erect apiculate lobes 1–1·6 mm. long ; midveins of petals distinct with ± long hairs, corolla otherwise densely puberulous outside, glabrous inside. Stamens with ovate shortly ciliate filament-bases ; anthers 1·3–2 mm. long. Ovary 3-locular ; style much shorter than the corolla-tube, 3-lobed, hairy at the base. Capsule 3-locular, dehiscing by basal valves. Seeds numerous, elliptic-oblong in outline, compressed, ± 0·6 mm. long, almost smooth, yellowish brown.

Kenya. Masai District: Ngong Hills, 28 Dec. 1954, *Bally* 9889 ! & 30 Nov. 1966, *Archer* 528 ! & 29 Nov. 1967, *Agnew* 9681 !
Distr. **K6** ; not known elsewhere
Hab. Upland grassland ; 2150–2430 m.

Syn. *C. sp. A* sensu Agnew, U.K.W.F. : 509 (1974)

Note. *C. edulis* and *C. keniensis* also differ in their chromosome numbers. They have 2n = 56 and 2n = 54 respectively.

INDEX TO CAMPANULACEAE

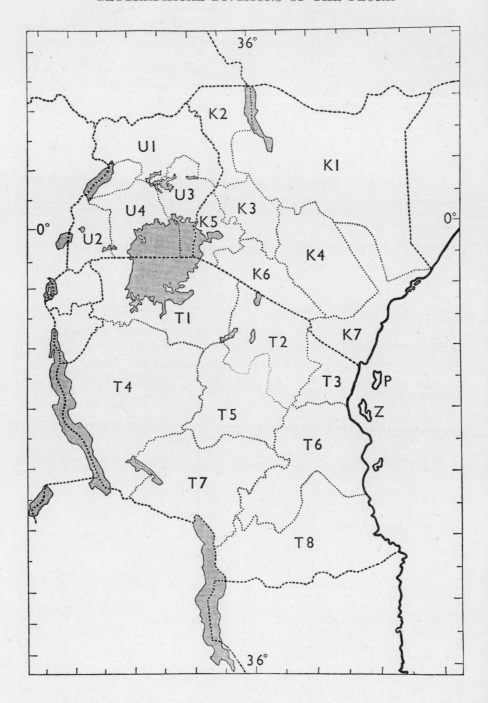